Synthesis and Application of Nano- and Microdispersed Systems

Synthesis and Application of Nano- and Microdispersed Systems

Editors

Denis Kuznetsov
Igor Burmistrov
Gopalu Karunakaran
Dmitry S. Muratov
Yudin Andrey

MDPI • Basel • Beijing • Wuhan • Barcelona • Belgrade • Manchester • Tokyo • Cluj • Tianjin

Editors

Denis Kuznetsov
University of Science and Technology "MISiS"
Russia

Igor Burmistrov
Saratov State Technical University
Russia

Gopalu Karunakaran
Seoul National University of Science and Technology (Seoul Tech)
Korea

Dmitry S. Muratov
University of Science and Technology "MISiS"
Russia

Yudin Andrey
University of Science and Technology "MISiS"
Russia

Editorial Office
MDPI
St. Alban-Anlage 66
4052 Basel, Switzerland

This is a reprint of articles from the Special Issue published online in the open access journal *Processes* (ISSN 2227-9717) (available at: https://www.mdpi.com/journal/processes/special_issues/nano_microdispersed).

For citation purposes, cite each article independently as indicated on the article page online and as indicated below:

LastName, A.A.; LastName, B.B.; LastName, C.C. Article Title. *Journal Name* **Year**, *Volume Number*, Page Range.

ISBN 978-3-0365-3691-0 (Hbk)
ISBN 978-3-0365-3692-7 (PDF)

Contents

About the Editors

Denis Kuznetsov is Head of the Department of Functional Nanosystems, National University of Science and Technology "MISiS", Leninsky prospect 4, 119049 Moscow, Russia. His research interests include nanomaterials, composites, and high-energy materials).

Igor Burmistrov is working as a Researcher in the Department of Physics, Saratov State Technical University, Politekhnicheskaya Ulitsa, 77, 410054 Saratov, Saratov Oblast, Russia. His research interests include nanomaterials, composites, and oxide materials for energy applications.

Gopalu Karunakaran is working as Research Professor and Brain Pool Fellow in the Biosensor Research Institute, Department of Fine Chemistry, Seoul National University of Science and Technology (Seoul Tech), Gongneung-ro 232, Nowon-gu, Seoul 01811, Republic of Korea. His research interests include nanomaterials and advanced materials for biomedical and energy applications.

Dmitry S. Muratov is working as a Researcher in the Department of Functional Nanosystems, National University of Science and Technology "MISiS", Leninsky prospect 4, 119049 Moscow, Russia. His research interests include nanocomposites and nanometal oxides for high temperature and energy applications.

Yudin Andrey is working as a Researcher at the Department of Functional Nanosystems, National University of Science and Technology "MISiS", Leninsky prospect 4, 119049 Moscow, Russia. His research interests include nanoaerosols and nanocomposites for energy applications.

Editorial

Special Issue on "Synthesis and Application of Nano- and Microdispersed Systems"

Denis Kuznetsov [1,*] and Gopalu Karunakaran [2,*]

1 Department of Functional Nanosystems and High-Temperature Materials, National University of Science and Technology "MISiS", Leninskiy Pr. 4, 119049 Moscow, Russia
2 Institute for Applied Chemistry, Department of Fine Chemistry, Seoul National University of Science and Technology (Seoultech), 232 Gongneung-ro, Nowon-gu, Seoul 01811, Korea
* Correspondence: dk@misis.ru (D.K.); karunakarang5@seoultech.ac.kr (G.K.)

With numerous advancements, nano- and microdispersed systems are rapidly increasing worldwide. [1]. This particularly includes nanoparticles and nanosystems. Generally, nano is defined as anything on the nanoscale range, from 1 to 100 nm [1]. However, larger sizes, from 100 to 500 nm, are also available for use. Based on the size and surface properties of nanoparticles, application variation occurs [2]. Major changes with respect to particle size and surface properties can lead to major changes in surface plasmon resonance, catalytic variation, magnetic properties, optical properties, and biological activities. Hence, it is essential to tune the structure of nanoparticles and their surface properties for better applications.

There are several approaches to nanoparticles' synthesis that researchers are working on at present. For example, the hydrothermal method, in which the synthesis of nanomaterial takes place at high pressure with high-temperature maintenance in a closed container, produces nanoparticles with a unique surface area and size. In addition, co-precipitation is also a common approach for the synthesis of nanoparticles. In this approach, two different solutes are precipitated together to form a unique and desired nanoparticle. This procedure requires a beaker, stirrers with magnetic bars, and the required solutes and solvents. Another approach to nanoparticle synthesis is chemical reduction. In this approach, the salt precursors are reduced with the reducing agents to produce metal and metal oxide nanoparticles for the desired applications. Furthermore, more recently, organic-microemulsion-based synthesis has been improving, in which different combinations of water, surfactant, and oils are used for the synthesis of nanoparticles. Moreover, a nanoparticle synthesis that is mediated by seeding is also performed, in which the seed nanoparticles are mixed with the salt precursors to form the desired nanoparticles. Hence, a variety of synthesis procedures are available, and each day, there are further advancements in the synthesis methods. Hence, this issue was planned to showcase the recent advancements in the synthesis of nanoparticles.

Furthermore, the applications of nanoparticles are included in this issue because the day-to-day applications of nanoparticles are increasing. Nanoparticles are used in wastewater treatments, sensors, electrode materials for lithium ion and sodium ion batteries, as super-capacitors, piezoelectric materials, optical applications, disease diagnosis materials and catalytic materials, in fuel cell applications, memory devices, antenna cores, pain industries, computer chips, and for biomedical applications. Thus, using this application knowledge, we aimed to build a research theme for the application of nanoparticles. Hence, we welcome the valuable contributions of all our researchers, in the form of original research papers, communications, and review papers on the synthesis and application of nanoparticles, using different key words, such as nanosystems, nanopowder, emulsion, and process.

We received a large number of papers for our Special Issue, which were peer-reviewed by experts in their respective fields. After a rigorous review process, ten original papers

Citation: Kuznetsov, D.; Karunakaran, G. Special Issue on "Synthesis and Application of Nano- and Microdispersed Systems". *Processes* **2022**, *10*, 608. https://doi.org/10.3390/pr10030608

Received: 25 February 2022
Accepted: 11 March 2022
Published: 21 March 2022

and one communication were published in this Special Issue. For example, Anastasia Yakusheva et al. described the synthesis of water-soluble carbon quantum dots for the detection of copper ions in aqueous media [1]. The carbon quantum dots were found to be very effective for the detection of copper in the aqueous medium. Anna A. Ulyankina et al. [3] synthesized TiO_2 nanoparticles with an electrochemical method using a pulse-alternating current for the photocatalytic oxidation of 5-hydroxymethylfurfural (HMF) to 2,5-diformylfuran (DFF). The TiO_2 nanoparticles were found to have higher DFF selectivity in methanol, of up to 33% when compared to the commercially available materials. Mai Ngoc Tuan Anh et al. [4] produced silver nanodecahedrons using photochemical synthesis for surface-enhanced Raman scattering (SERS) properties and applications. It was observed that the synthesized silver nanodecahedrons exhibited stronger SERS properties after 48 h of Light-Emitting Diode (LED) radiation. N. Van Minh et al. [5] synthesized spark plasma-sintered cobalt materials for enhanced mechanical properties. It was found that a high bending strength, microhardness, and relative density enhancement were achieved with the nanomodification of spark plasma-sintered cobalt materials.

Hence, this Special Issue on nano- and microdispersed systems is very useful to any researchers looking for new synthesis and applications, such as sensors, wastewater treatment, and chemical conversion. This Special Issue will also be a valuable reference point for future research in this field. From our Guest Editor team board, we congratulate and thank all the authors who made valuable contributions to this issue. We also sincerely thank all the expert members and reviewers who spent their valuable time and effort reviewing all the papers. We would also like to thank the editorial team and all the technical staff of *Processes* for their efforts.

Author Contributions: D.K. and G.K. writing—review and editing. All authors have read and agreed to the published version of the manuscript.

Funding: This research received no external funding.

Institutional Review Board Statement: Not applicable.

Informed Consent Statement: Not applicable.

Data Availability Statement: Not applicable.

Conflicts of Interest: The authors declare no conflict of interest.

References

1. Yakusheva, A.; Muratov, D.S.; Arkhipov, D.; Karunakaran, G.; Eremin, S.A.; Kuznetsov, D. Water-Soluble Carbon Quantum Dots Modified by Amino Groups for Polarization Fluorescence Detection of Copper (II) Ion in Aqueous Media. *Processes* **2020**, *8*, 1573. [CrossRef]
2. Ermolenko, A.; Vikulova, M.; Shevelev, A.; Mastalygina, E.; Ogbuna Offor, P.; Konyukhov, Y.; Razinov, A.; Gorokhovsky, A.; Burmistrov, I. Sorbent Based on Polyvinyl Butyral and Potassium Polytitanate for Purifying Wastewater from Heavy Metal Ions. *Processes* **2020**, *8*, 690. [CrossRef]
3. Ulyankina, A.; Mitchenko, S.; Smirnova, N. Selective Photocatalytic Oxidation of 5-HMF in Water over Electrochemically Synthesized TiO_2 Nanoparticles. *Processes* **2020**, *8*, 647. [CrossRef]
4. Tuan Anh, M.N.; Nguyen, D.T.D.; Ke Thanh, N.V.; Phuong Phong, N.T.; Nguyen, D.H.; Nguyen-Le, M.-T. Photochemical Synthesis of Silver Nanodecahedrons under Blue LED Irradiation and Their SERS Activity. *Processes* **2020**, *8*, 292. [CrossRef]
5. Nguyen, V.M.; Khanna, R.; Konyukhov, Y.; Nguyen, T.H.; Burmistrov, I.; Levina, V.; Golov, I.; Karunakaran, G. Spark Plasma Sintering of Cobalt Powders in Conjunction with High Energy Mechanical Treatment and Nanomodification. *Processes* **2020**, *8*, 627. [CrossRef]

Article

Water-Soluble Carbon Quantum Dots Modified by Amino Groups for Polarization Fluorescence Detection of Copper (II) Ion in Aqueous Media

Anastasia Yakusheva [1,*], Dmitry S. Muratov [1], Dmitry Arkhipov [1], Gopalu Karunakaran [2], Sergei A. Eremin [3] and Denis Kuznetsov [1]

1 Department of Functional Nanosystems and High-Temperature Materials, National University of Science and Technology MISIS, Leninsky prospect 4, 119049 Moscow, Russia; muratov@misis.ru (D.S.M.); hipa2010@yandex.ru (D.A.); dk@misis.ru (D.K.)

2 Biosensor Research Institute, Department of Fine Chemistry, Seoul National University of Science and Technology (SeoulTech), Gongneung-ro 232, Nowon-gu, Seoul 01811, Korea; karunakarang5@gmail.com

3 Faculty of Chemistry, M. V. Lomonosov Moscow State University, Lenin's mountains, 119991 Moscow, Russia; saeremin@gmail.com

* Correspondence: yakusheva.as@misis.ru; Tel.: +7-909-669-30-81

Received: 6 November 2020; Accepted: 21 November 2020; Published: 29 November 2020

Abstract: Industrialization is serious for changing the environment and natural water composition, especially near cities and manufacturing areas. Logically, the new ultrasensitive technology for precise control of the quality and quantity of water sources is needed. Herein, an innovative method of polarization fluorescence analysis (FPA) was developed to measure the concentration of heavy metals in water. The approach was successfully applied for precise tests with reduced analysis time and increased measurement efficiency among laboratory methods. Based on this work, the investigations established the new type of carbon quantum dots (CQDs) with controllable fluorescence properties and functionalized amino—groups, which is appropriate for FPA. The parameters of one and two-step microwave synthesis routes are adjusted wavelength and fluorescence intensity of CQDs. Finally, under optimized conditions, the FPA is showed the detection of copper (2+) cations in water samples below European Union standard (2 mg/L). Moreover, in comparison with fluorescence quenching, polarization fluorescence is proved as a convenient, simple, and rapid test method for effective water safety analysis.

Keywords: carbon quantum dots; polarization fluorescent analysis; fluorescence properties; copper cation; water samples

1. Introduction

Recently, heavy metals have been shown to be a severe problem due to active industrialization. The problem of environmental pollution is still the main attraction to enter eco-friendly technology. A promising solution is needed for the present tests of toxic chemical elements, compounds, and active ions in the environment sphere. The new alternative technology demands the new investigation and enhancement for keeping up with the changing world [1]. Commonly, many reasons are compared for appropriate research. Measurements will have identified a wide range of chemical compounds, simultaneously with high performance, cost-effective, eco-friendly probe preparation, and elementary experiment stages have not been [SL1] solved completely [2–4].

In the past few decades, the quantitative and qualitative laboratory methods include chromatography, spectrophotometry, solid-phase extraction with atomic absorption spectroscopy, and atomic emission spectroscopy, and mass-spectrometry with inductively coupled plasma for determining a heavy metal in water probe are generally used [5–10]. However, the methodologies are not so much better for the large-scale laboratory measurements with the growing demand. One of the strategies is to apply functional sensors, which are dedicated to a new measurement method [11]. From this side, quantum dots as optical nanosensors based on zinc, selenium, tellurium, and other chemical element have a greater interest in the studies. Carbon dots are one of the best materials for eco-friendly application. Today, particles are actively investigated as sensors for determining the concentration of chemical compounds in aqueous media. There are so many different types of inorganic nanoparticles with significant drawbacks: expensive precursors for synthesis, insolubility in water, and toxic to the environment [12].

A new promising class of zero-dimensional carbon materials was evaluated as sensors from bio and eco-compatible precursors with fine optical properties [13,14]. Over the past few decades, there has been sustained research activity on carbon quantum dots (CQDs) with a size up to 10 nm [15,16] due to stable and sensitively fluorescent properties [17]. A number of various types of CQDs are still getting bigger over time, but the synthesis stage is common for all [18]. The synthesis of CQDs is carried out in three separate routes: carbonization, passivation, and surface functionalization. The temperature of carbonization is a very sensitive parameter above 300 degrees for the decomposition of carbon-containing precursor [19]. For example, a plant material, such as fruit and nut peels, berries, and juices, have shown the retention of specific functional groups [20] only up to 250 °C on particle surface [21]. Moreover, the fluorescence spectrum and quantum yield of CQDs significantly depend on the carbon and nitrogen ratio in the particle and suffer from inconsistent precursors. Consequently, the preferable synthesis technique for adjusting the first stage is to use chemical compounds instead of natural precursors [22]. Briefly, sulfur, nitrogen, and oxygen as inclusions are mutually involved in nanoparticles for increased quantum yield [23]. Functional groups on the nanoparticle surface are determined by practical application [24,25] in waste-water sensors technology [26]. The type and number of groups influence optical properties [27]. For example, surface adsorption of harmful chemical compounds leads to a strong fluorescence quenching [28,29]. The impurities and functional groups are precisely coordinated contaminant molecules on the surface for excellent sensor properties. While a variety of synthesis routes have been suggested, all its demanding hard-controllable morphology optimization, reduction of surface defects, trap sites, as a result, enhance the fluorescence intensity of passivated carbon nanoparticles.

Usually, the detection of heavy metals in samples is under consideration for concentration cations in aqueous containing probe [30,31]. The measurements based on fluorescent, colorimetric sensors [32,33], and visual detection [34] have many problems with pH tolerance, experiment time, and solvent conditions of actual samples [35].

Herein, polarization fluorescence analysis (FPA) is represented as a new alternative optical measurement. In this perspective, FPA establishes the fluorophore polarization effect, which depends on surface absorption and sample viscosity. The parameters include more proven information about sensitive chemical coordination near the double electron layer of the CQD surface. The measurement data of FPA, in comparison with fluorescence quenching methods, confirm the effectiveness of CQDs nano-sensors to determine copper cations in water.

2. Materials and Methods

2.1. Reagents and Materials

The precursors used in the synthesis: citric acid (99.8%), ascorbic acid (99.9%), ammonium dihydrogen phosphate (99.0%), and Trilon-B (EDTA-Na_2: disodium salt of ethylenediaminetetraacetic acid) (99.5%). The analytical standards of Fe, Co, Ni, Cu, Zn, Ag, Au, Hg, Pb, and Sn, in the range from 1×10^{-5} to 1×10^{3} mg/L concentration apply for confirming selective detection of the copper cation in water.

2.2. Instruments

Herein, the SUPRA MWS-1803MW microwave oven (Hangzhou, China) was used for the synthesis procedure under 700 W microwave irradiation. The size distribution and morphology of CQDs were characterized by transmission electron microscopy JEOL JEM-1400 (Tokyo, Japan). The size measurements were provided on Malvern Zetasize Nano ZS particle size and particle zeta-potential analyzer ZEN3600 (Malvern Panalytical Ltd, Malvern, UK). The optical properties of CQDs were studied using the Agilent Technologies Cary Eclipse Fluorescence spectrophotometer (Santa Clara, CA, USA) and Thermo Scientific HeλiosHeλios α spectrophotometer (Waltham, MA, USA). IR-spectra of CQDs were obtained from Thermo Scientific Spectrometer Thermo NICOLET 380 (Waltham, MA, USA) to identify the functional groups. The express fluorescence polarization and intensity tests were conducted at portable fluorimeter Sentry-100 (Waukesha, WI, USA).

2.3. Synthesis of CQDs

Firstly, the dry precursor's mixtures were used for the first stage: CQD 1 (0.044 mol of citric acid and 0.022 mol of ammonium dihydrogen phosphate), CQD 2 (0.044 mol of ascorbic acid and 0.022 mol of ammonium dihydrogen phosphate), and CQD 3 (0.044 mol of citric acid, 0.022 mol of ammonium dihydrogen phosphate and 0.004 mol of ethylenediaminetetraacetic acid disodium salt) were heated for 1 min. The fluorescence wavelength and intensity were investigated by adding disodium salt of ethylenediaminetetraacetic acid in the following amounts: 0.004, 0.005, 0.006, 0.007, and 0.008 mol.

For the two-step synthesis route, the 0.044 mol of ascorbic acid and 0.022 mol of ammonium dihydrogen phosphate were irradiated for 1 min, and then 0.004 mol of ethylenediaminetetraacetic acid disodium salt was added to continue reaction in 30 s for CQD 4. Additionally, the variety amount 0.004, 0.005, 0.006, 0.007, and 0.008 mol was added in the second step.

In the experiment, the sample in the crucible was cooled to room temperature and suspended in 3×10^{-2} L of water to collect the dark brown dispersion. The next required step is filtering through a Millex-LG 0.2 µm IC filter cartridge (Millipore, Burlington, MA, USA). The 1×10^{-2} L of CQDs in a test tube were centrifuged for 30 min on 12.4 g acceleration (14,500 rpm) to remove large particles. More synthesis parameters are shown in Table 1.

Table 1. Ratio of precursors and time of microwave irradiation for carbon quantum dots (CQDs) sample.

S.No.	Sample Code	Citric Acid, mol	Ascorbic Acid, mol	Ammonium Dihydrogen Phosphate, mol	1st Step Microvalve Irradiation, s	Trilon-B (EDTA-Na_2), mol	2nd Step Microvalve Irradiation, s
1.	CQD 1	0.044	-	0.022	60	-	-
2.	CQD 2	-	0.044	0.022	60	-	-
3.	CQD 3_1	0.044	-	0.022	60	0.004	-
4.	CQD 3_2	0.044	-	0.022	60	0.005	-
5.	CQD 3_3	0.044	-	0.022	60	0.006	-
6.	CQD 3_4	0.044	-	0.022	60	0.007	-
7.	CQD 3_5	0.044	-	0.022	60	0.008	-
8.	CQD 4_1	0.044	-	0.022	60	0.004	30
9.	CQD 4_2	0.044	-	0.022	60	0.005	30
10.	CQD 4_3	0.044	-	0.022	60	0.006	30
11.	CQD 4_4	0.044	-	0.022	60	0.007	30
12.	CQD 4_5	0.044	-	0.022	60	0.008	30

2.4. Detection of Metal Ions

The FPA is establishing a technique of measurements that considers surface absorption and sample viscosity more informative in comparison with fluorescence intensity. According to the results, CQD with constant fluorescence emission was preferable for measurement on portable fluorimeter

SENTRY-100. The optical system of fluorimeter is poorer than laboratory equipment due to the optical detector on 535 nm wavelength emission.

In experiments, the optical parameters of CQDs calculate for concentration from 0.1 mg/L to 20 mg/L in dispersion. In addition, the fluorescence quenching confirms on Agilent Technologies Cary Eclipse Fluorescence spectrophotometer. The results fluorescence intensity (FI) of CQD with different positions of the maximum emission spectra were chosen as the most appropriate particles and compared with fluorescence intensity data on SENTRY-100 in Figure 1.

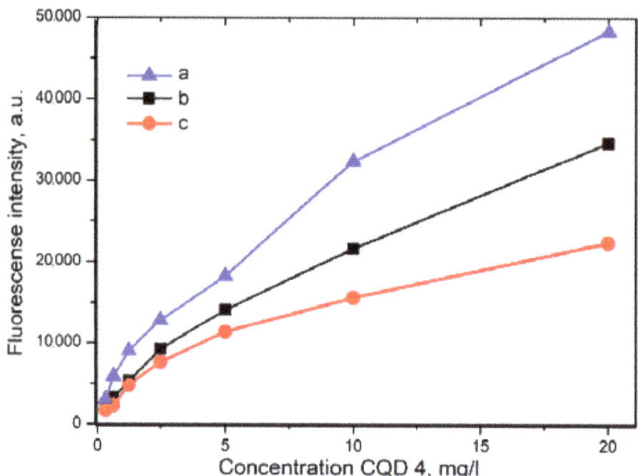

Figure 1. Fluorescence intensity of CQD 4 depends on the concentration in deionized water. The emission peak is 530 nm (**a**), 560 nm (**b**), and 500 nm (**c**), respectively.

As reported before, the fluorescence intensity of CQD 4 was better for samples with the position of fluorescence peak at 530 nm due to properties of CQD (530 nm) are more closest to the SENTRY-100 optical system, then 500 nm (Figure 1c) and 560 nm (Figure 1b). Moreover, the higher fluorescence intensity may apply to reduce the concentration of CQDs in the measurement cuvette and also increase the experiment sensitivity.

In this case, the CQD dispersion (the position of maximum fluorescence at 530 nm) with a final concentration at 6.8 mg/L was designed, which corresponds to recommended fluorescence intensity 24,000 a.u. to achieve the maximum accuracy for FI and FP. The technique of metal cations concentration measuring was studied in standard cations solutions of Fe, Co, Ni, Cu, Zn, Ag, Au, Hg, Pb, and Sn. The concentration of metal cations was used from 10^{-5} mg/L to 10^{3} mg/L and added in a volume of 2×10^{-5} L in mL CQD's sample.

3. Results and Discussion

3.1. Characterization of CQD

In the present study, the better alternative microwave assistance synthesis of amino-functionalized CQDs present. The order to evaluate the optical properties depend on the synthesis route [36], the influence of precursors (a type of carbon source) was identified in Figure 2.

The core of fluorescence CQD 1 and CQD 2 with a varied source for carbonization stage correspond with acid sources and position of absorption (ABS) peak. Briefly, in the equal amount of citric acid or ascorbic acid, the ABS peak of CQD 1 at 250 nm, and the peak of CQD 2 is attributed to the right on 90 nm. The evolution between the Stokes shift of the first and second type of nanoparticles is slightly different. The fluorescence (FL) peak excited at a maximum of ABS wavelength shown in Figure 2B at 410 nm and 500 nm for CQD 1 and CQD 2, respectively. Optical curves have been in equal intensity of excitation and emission.

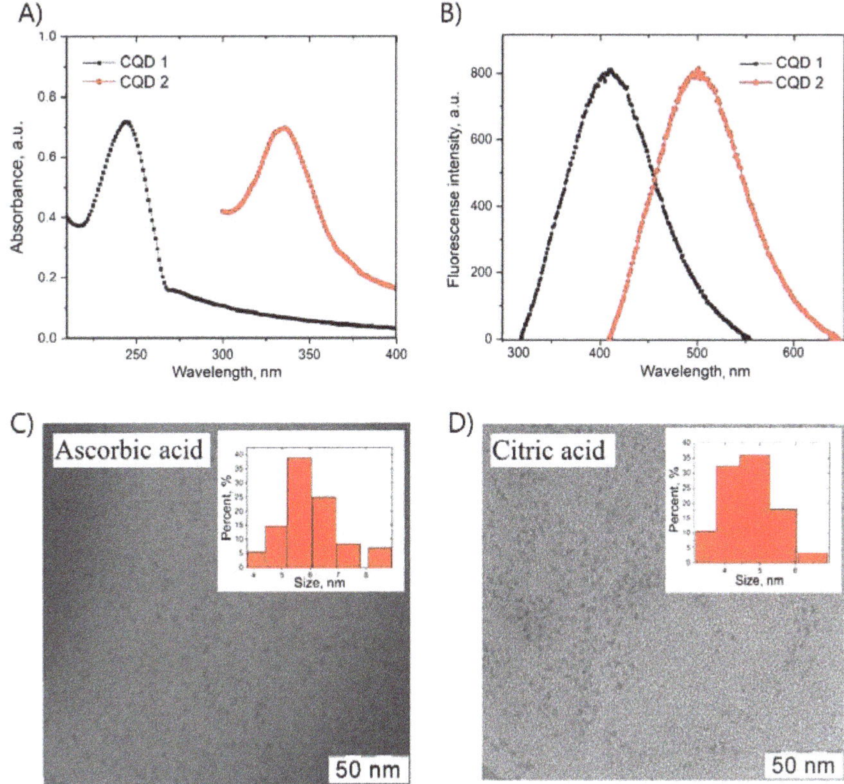

Figure 2. Characterizations carbon core of CQD 1, CQD 2, (**A**) ABS curve CQDs from ascorbic or citric acid as carbon source, (**B**) FL spectra of CQD 1 and CQD 2, corresponding with ABS spectra, (**C**) TEM image and size distribution of CQD 1, and (**D**) TEM image and size distribution of CQD 2.

The morphology of nanoparticles and size distribution assessment provide from the TEM image. Pictures showed similar quasi-spherical CQDs with size near 5 nm at both synthesis routes. Noticeably, the average size for each type of dot is not correlating to FL peak position due to relative Stock shift at ABS spectra attribute to the structure of the carbon bond in precursors [37].

3.2. Fluorescence Properties of CQD

In this case, the carbon source determined the position of maximum fluorescence as a limit factor for CQDs design. The acids for CQD were saved in a molar ratio to ammonium dihydrogen phosphate (oxidizer with phosphorus and nitrogen inclusions). The constant ratio supports the uniform fluorescent core of carbon dots in the synthesis condition. Secondly, a similar effect was confirmed at the amount of EDTA-Na_2, if comparing CQD 3 and CQD 4, which was shown in spectra in Figure 3A–D, respectively. In comparison, the ability of two-steps microwave irradiation to improve fluorescence properties proven here.

As exhibited in spectra CQD 3 and CQD 4 in Figure 3A,C, the two-steps microwave irradiation plays a key role in surface energy formation. In particular, the excitation in Figure 3C has attracted attention to absorbance shift.

Moreover, in Figure 3B,D, the FI of CQD 4 is intensive twice more than CQD 3 and shows the green and yellow colored light. Briefly, the CQD 3 represented a maximum emission wavelength of 535 nm under 380 nm of exciting wavelength and all emission spectra in the 500–535 nm range. Opposite the

CQD 4 shows the broad range of emission from 510 to 570 nm and a maximum emission wavelength of 570 nm under 400 nm of exciting wavelength. The key factor of the emission dependence is associated with the coordinating center in EDTA-Na_2 [38], which is safe for the structure and evenly distributed on the particle surface. Consequently, the two-step microwave synthesis CQD 4 was shown to have a better ability for fluorescent intensity, which can be controlled by changing the type and number of precursors for functionalization.

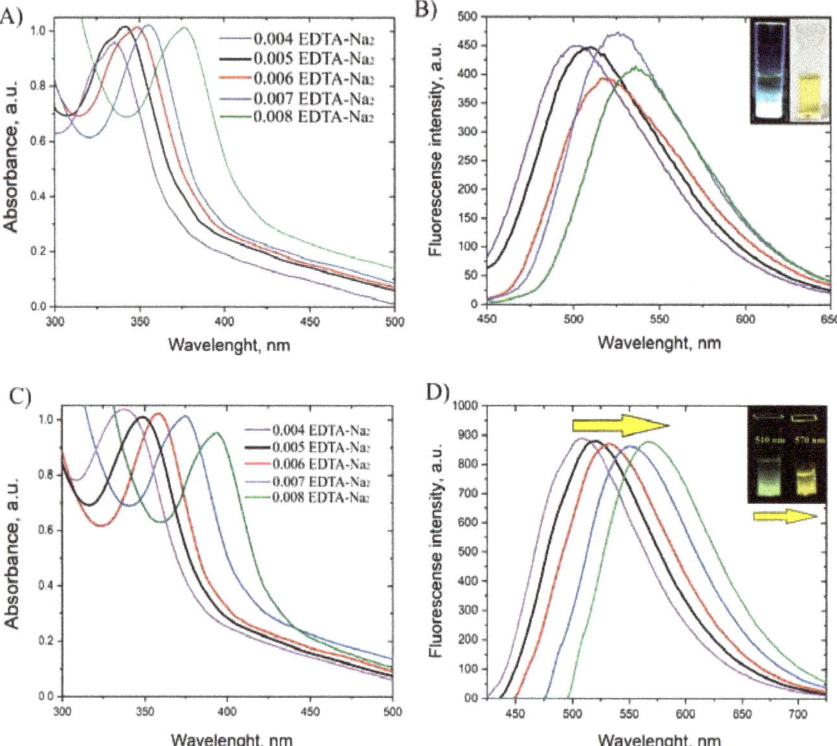

Figure 3. (**A**) Absorbance spectra of CQD 3 with a different mol EDTA-Na_2 (0.004, 0.005, 0.00. In this case, the CQD dispersion (the position of maximum fluorescence at 530 nm) with a final concentration at 0.68 mg/L was designed, 0.007, 0.008), (**B**) emission spectra of CQD 3, respectively, (**C**) absorbance spectra of CQD 4 with a different mol EDTA-Na_2 (0.004, 0.005, 0.006, 0.007, 0.008), (**D**) emission spectra of CQD 4.

Size Distribution of CQD 1, 3 and 4

To explore fluorescence behavior, secondly, the size distribution of CQD was measured by a dynamic light scattering method in Figure 4.

The size distribution exhibited the structure of the nanoparticles and affected to fluorescence properties of quantum dots [39]. Herein, the curves of size distribution in Figure 3A correspond with three types of CQD 1, CQD 3.3, and CQD 4.3 particles, respectively. These are attributed to the core (CQD 1), the average size was near 3–4 nm. If EDTA-Na_2 is added during the synthesis rote, the average size increases to 6–7 nm for the one-step process and 8–9 nm for the two-step process, as shown in Figure 3B, due to consecutive carbonization of precursors. The results show that the average size of CQDs was effectively changed by steps of synthesis (in particular, the time of microwave irradiation) [40].

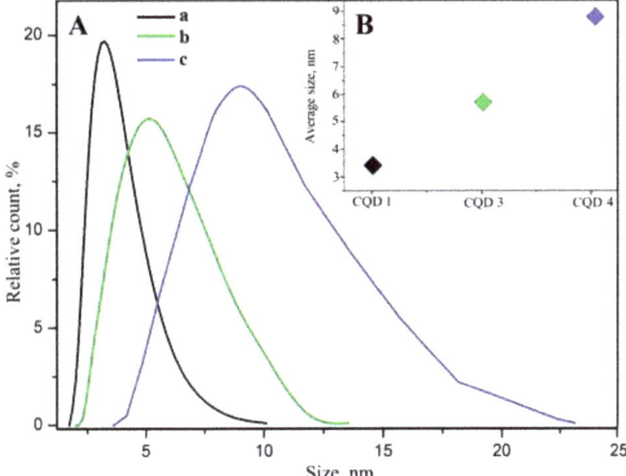

Figure 4. (**A**) Size distribution of CQD 1, CQD 3, and CQD 4. (**B**) The average size of CQD 1, CQD 3 and CQD 4.

3.3. FI Quench of CQD by Addition of Metal Ions

The technique of metal cations concentration measuring by CQD 4 with appropriate controllable optical properties was studied in standard cations solutions of Fe, Co, Ni, Cu, Zn, Ag, Au, Hg, Pb, and Sn. The results of fluorescence quenching for metal salt solutions were presented in Figure 5.

The bar charts in Figure 5A show that 5×10^{-5} L all-metal salt solution with 10^{-3} L CQD 4 dispersion close to fluorescence quenching to pure CQDs solution. In opposite to previous metals, fluorescence intensity dramatically decreased after Cu^{2+} cations-coordinating reaction at CQD 4. In particular, the result established reaction activity between nanoparticle and cupper cations [41]. As shown in Figure 5B,D, cupper cations can effectively quench the fluorescence of CQD 4 from 1×10^{-4} mg/mL to 1 g/mL. During the experiment, the accuracy of measurements is confirming by the added-found approach. Fluorescence quenching demonstrated an interaction between the copper cation and CQD.

First of all, the exploration of chemistry starts with FTIR analysis. The method was used as an indicator to optimize measurements of Cu^{2+} in water samples. The FTIR spectra in Figure 5C. are shown the functional groups on CQD 4 before adding the Cu^{2+} solution Figure 5Ca and after Figure 5Cb. The CQD 4 reactions on the surface are characterizing by the peaks at 3420 cm^{-1} and 1420 cm^{-1} due to the change in line intensity of hydroxyl groups. A peak at 1700 cm^{-1} is attributed to the carboxyl group, which is retained from citric acid. The peak at 1650 cm^{-1} was correlated with the valence C = N bond and amino-groups on the surface of nanoparticles correspond with the peak of 1095 cm^{-1}. Carbon- carbon bonds, like C − C and C = C, are reflected by peaks of 1360 cm^{-11}, 1230 cm^{-1}. In summary, the IR-spectra, the hydroxyl, carboxyl, and amino groups on the CQDs surface were found. Copper was causing an intensity change in the spectrum, which depend on the chemical bonds in functional groups. The greatest change corresponds with Cu^{2+} and amino group reaction via non-fluorescence complexes formation [42,43].

The concentration of copper ions was investigated in both: the FPA and FI quenching method, in comparison by Sentry-100 measurements. The data of fluorescence polarization correlated with the average angular displacement of the fluorophore, which occurs absorption to emission photon. This value depends on the rotational diffusion of a fluorophore, which is determined by the viscosity of the sample, temperature, and particle size. When viscosity and temperature are constant, the fluorescence polarization shows particle size changes due to copper is joined to the CQD surface. The ratio FI, FP, and concentration of copper cations are presented in Figure 6.

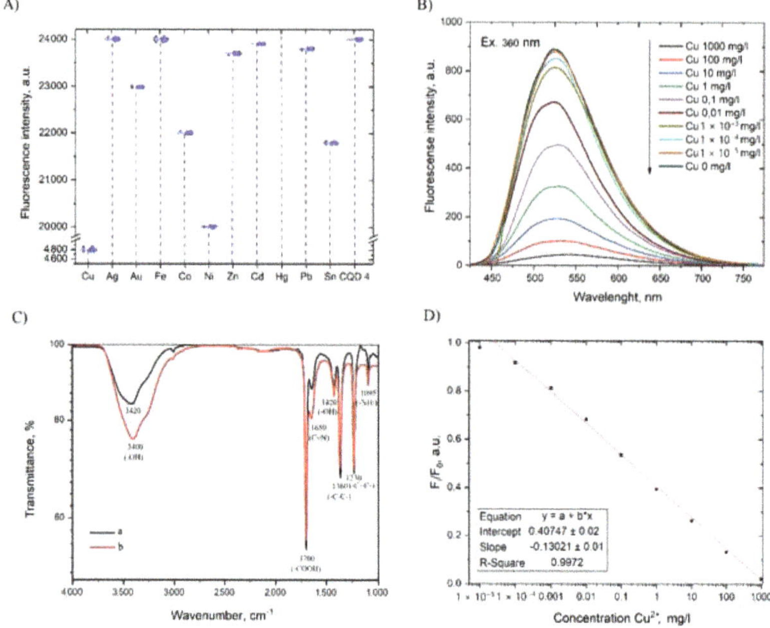

Figure 5. (**A**) The fluorescence quenching effect of CQD 4 with metal cation in water, (**B**) Concentration dependence for the fluorescence quenching of PL CQD 4 by Cu^{2+}, (**C**) FTIR spectra of CQD 4 before (a) and after (b) coordinated Cu^{2+} cations, and (**D**) The linear relationship between the fluorescence intensity CQD 4 and the Cu^{2+} concentration.

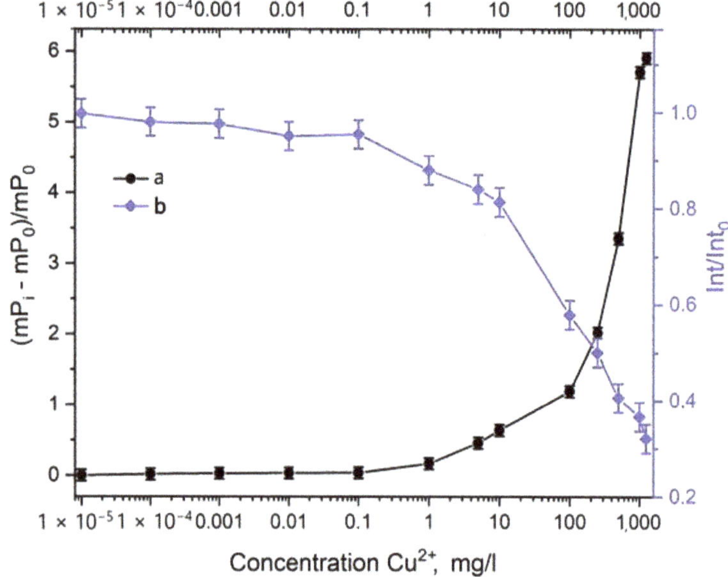

Figure 6. Relative changes between fluorescence polarisation (**a**) and fluorescence intensity (**b**) of CQD 4 sample with different copper (2+) cations concentration.

For FPA rapid analysis (Figure 6a) low detection limit of Cu^{2+} concentration is 1 mg/L, which is comparable to fluorescence intensity (Figure 6b) measurements from outside laboratory location [44]. The method of measuring the fluorescence polarization allows determining concentrations, which are compared with the fluorescence intensity method, but the errors show a high accuracy level. In addition, functional groups react with copper ions, hence, the average size of carbon nanoparticles was increased. The relative change in polarization values shows a fivefold data increase, opposite to fluorescence intensity is limited to a range of measured concentrations. At the same, the relative measurement error is reduced when using the method of FPA, which is more suitable for the determination of copper.

In particular, a calibration graph for FPA of Cu^{2+} concentration and fluorescence polarization of CQD 4 is presented in Figure 7.

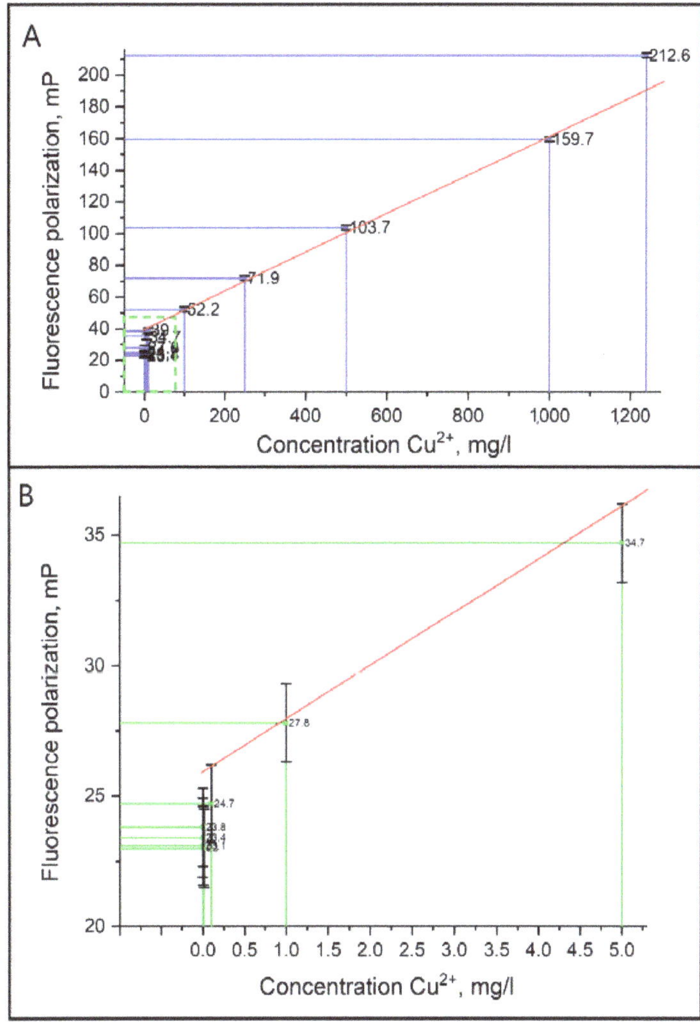

Figure 7. Calibration graph (A) CQD 4 fluorescence polarization values and copper (2+) cation concentrations in the water sample from 1 mg/L to 1240 mg/L, (B) Up-scale graph for 0 mg/ml to 5 mg/ml copper (2+) cation concentrations

A detection of limit in a water sample (7.7 μmol/L) investigated via the added-found method when the linear relationship between the quantities from 1 mg/L to 1g/L (2×10^{-5} L added in a sample) is constant. For comparative, standards of safe copper concentration in natural water 2 mg/L (WHO, EU). Consequently, the FPA can be used to determine the safety and ecological state of water resources.

4. Conclusions

Recent studies have shown efficient two-step microwave-assisted synthesis of CQD. Nanoparticles with improved optical properties and fluorescence maximum from 510 to 570 nm get via changing the precursor amount at the second step of the approach. The amino groups on the surface of CQD 4 allowed the use of carbon nanoparticles for selective research of copper cations in aqueous samples from 1 mg/L to 1 g/L. Confirming result is based on polarization fluorescence analysis, which admitted to the measurement of a hundred simultaneous water intake samples. One of the advantages of the method is the measurement ability for hard-to-reach or remote water sources. In conclusion, the time to time screening of water safety and assessment of copper cations at the appropriate level, which satisfactorily meets the limit of Cu^{2+} ions in drinking water (2 mg/L EU standard), was proven by FPA based on CQD 4.

Author Contributions: Conceptualization, A.Y., D.K., S.A.E.; Methodology, A.Y., S.A.E.; Software, D.S.M.; Validation, G.K., S.A.E., D.S.M.; Formal Analysis, A.Y.; Investigation, A.Y.; Resources, D.K., S.A.E., D.A.; Data Curation, A.Y.; Writing—Original Draft Preparation, A.Y., D.S.M.; Writing—Review and Editing, G.K.; Visualization, D.A.; Supervision, D.K.; Project Administration, D.K. Funding Acquisition, D.A. All authors have read and agreed to the published version of the manuscript.

Funding: The authors gratefully acknowledge the financial support of the Russian Foundation for Basic Research according to the research project No. 18-29-25051.

Conflicts of Interest: The authors declare no conflict of interest.

References

1. Mohankumar, K.; Hariharan, V.; Rao, N.P. Heavy Metal Contamination in Groundwater around Industrial Estate vs. Residential Areas in Coimbatore, India. *J. Clin. Diagn. Res.* **2016**, *10*, BC05–BC07. [CrossRef] [PubMed]
2. Yang, J.-Y.; Zhang, Y.; Wang, H.; Xu, Z.-L.; Eremin, S.A.; Shen, Y.-D.; Wu, Q.; Lei, H.; Sun, Y.-M. Development of fluorescence polarisation immunoassay for carbofuran in food and environmental water samples. *Food Agric. Immunol.* **2014**, *26*, 340–355. [CrossRef]
3. Zhang, H.; Yang, S.; De Ruyck, K.; Beloglazova, N.V.; Eremin, S.A.; De Saeger, S.; Zhang, S.; Shen, J.; Wang, Z. Fluorescence polarization assays for chemical contaminants in food and environmental analyses. *Trac Trends Anal. Chem.* **2019**, *114*, 293–313. [CrossRef]
4. McGillicuddy, N.; Nesterenko, E.P.; Nesterenko, P.N.; Jones, P.; Paull, B. Chelation ion chromatography of alkaline earth and transition metals a using monolithic silica column with bonded N-hydroxyethyliminodiacetic acid functional groups. *J. Chromatogr. A* **2013**, *1276*, 102–111. [CrossRef]
5. Lopes, F.S.; Junior, O.A.; Gutz, I.G.R. Fully electrochemical hyphenated flow system for preconcentration, cleanup, stripping, capillary electrophoresis with stacking and contactless conductivity detection of trace heavy metals. *Electrochem. Commun.* **2010**, *12*, 1387–1390. [CrossRef]
6. Wen, X.; Yang, Q.; Yan, Z.; Deng, Q. Determination of cadmium and copper in water and food samples by dispersive liquid–liquid microextraction combined with UV–vis spectrophotometry. *Microchem. J.* **2011**, *97*, 249–254. [CrossRef]
7. Ghanemi, K.; Nikpour, Y.; Omidvar, O.; Maryamabadi, A. Sulfur-nanoparticle-based method for separation and preconcentration of some heavy metals in marine samples prior to flame atomic absorption spectrometry determination. *Talanta* **2011**, *85*, 763–769. [CrossRef]
8. Li, W.; Simmons, P.; Shrader, D.; Herrman, T.J.; Uzuner, U. Microwave plasma-atomic emission spectroscopy as a tool for the determination of copper, iron, manganese and zinc in animal feed and fertilizer. *Talanta* **2013**, *112*, 43–48. [CrossRef]

9. Dai, B.; Cao, M.; Fang, G.; Liu, B.; Dong, X.; Pan, M.; Wang, S. Schiff base-chitosan grafted multiwalled carbon nanotubes as a novel solid-phase extraction adsorbent for determination of heavy metal by ICP-MS. *J. Hazard. Mater.* **2012**, *219*, 103–110. [CrossRef]
10. Das, P.; Ganguly, S.; Bose, M.; Mondal, S.; Das, A.; Banerjee, S.; Das, N.C. A simplistic approach to green future with eco-friendly luminescent carbon dots and their application to fluorescent nano-sensor 'turn-off' probe for selective sensing of copper ions. *Mater. Sci. Eng. C* **2017**, *75*, 1456–1464. [CrossRef]
11. Smith, D.S.; Eremin, S.A. Fluorescence polarization immunoassays and related methods for simple, high-throughput screening of small molecules. *Anal. Bioanal. Chem.* **2008**, *391*, 1499–1507. [CrossRef] [PubMed]
12. Hardman, R. A Toxicologic Review of Quantum Dots: Toxicity Depends on Physicochemical and Environmental Factors. *Environ. Health Perspect.* **2006**, *114*, 165–172. [CrossRef] [PubMed]
13. Zuo, P.; Lu, X.; Sun, Z.; Guo, Y.; He, H. A review on syntheses, properties, characterization and bioanalytical applications of fluorescent carbon dots. *Microchim. Acta* **2015**, *183*, 519–542. [CrossRef]
14. Simpson, A.; Pandey, R.; Chusuei, C.C.; Ghosh, K.; Patel, R.; Wanekaya, A.K. Fabrication characterization and potential applications of carbon nanoparticles in the detection of heavy metal ions in aqueous media. *Carbon* **2018**, *127*, 122–130. [CrossRef]
15. Al-Douri, Y.; Badi, N.; Voon, C.H. Synthesis of carbon-based quantum dots from starch extracts: Optical investigations. *Luminescence* **2018**, *33*, 260–266. [CrossRef]
16. Yuan, F.; Wang, Z.; Li, X.; Li, Y.; Tan, Z.; Fan, L.; Yang, S. Bright Multicolor Bandgap Fluorescent Carbon Quantum Dots for Electroluminescent Light-Emitting Diodes. *Adv. Mater.* **2017**, *29*, 1–6. [CrossRef]
17. Fan, Y.Z.; Zhang, Y.; Li, N.; Liu, S.G.; Liu, T.; Li, N.B.; Luo, H.Q. A facile synthesis of water-soluble carbon dots as a label-free fluorescent probe for rapid, selective and sensitive detection of picric acid. *Sens. Actuators B Chem.* **2017**, *240*, 949–955. [CrossRef]
18. Moreira, V.A.; Suarez, W.T.; Franco, M.D.O.K.; Neto, F.F.G. Eco-friendly synthesis of cuprizone-functionalized luminescent carbon dots and application as a sensor for the determination of copper(ii) in wastewater. *Anal. Methods* **2018**, *10*, 4570–4578. [CrossRef]
19. Xu, Y.; Tang, C.-J.; Huang, H.; Sun, C.-Q.; Zhang, Y.-K.; Ye, Q.-F.; Wang, A.-J. Green Synthesis of Fluorescent Carbon Quantum Dots for Detection of Hg^{2+}. *Chin. J. Anal. Chem.* **2014**, *42*, 1252–1258. [CrossRef]
20. Ma, X.; Dong, Y.; Sun, H.; Chen, N. Highly fluorescent carbon dots from peanut shells as potential probes for copper ion: The optimization and analysis of the synthetic process. *Mater. Today Chem.* **2017**, *5*, 1–10. [CrossRef]
21. LeCroy, G.E.; Yang, S.-T.; Yang, F.; Liu, Y.; Fernando, K.A.S.; Bunker, C.E.; Hu, Y.; Luo, P.G.; Sun, Y.-P. Functionalized carbon nanoparticles: Syntheses and applications in optical bioimaging and energy conversion. *Co-Ord. Chem. Rev.* **2016**, *320–321*, 66–81. [CrossRef]
22. Dai, Y.; Liu, Z.; Bai, Y.; Chen, Z.; Qin, J.; Feng, F. A novel highly fluorescent S, N, O co-doped carbon dots for biosensing and bioimaging of copper ions in live cells. *Rsc Adv.* **2018**, *8*, 42246–42252. [CrossRef]
23. Fernando, K.A.S.; Sahu, S.; Liu, Y.; Lewis, W.K.; Guliants, E.A.; Jafariyan, A.; Wang, P.; Bunker, C.E.; Sun, Y.-P. Carbon Quantum Dots and Applications in Photocatalytic Energy Conversion. *Acs Appl. Mater. Interfaces* **2015**, *7*, 8363–8376. [CrossRef] [PubMed]
24. He, G.; Shu, M.; Yang, Z.; Ma, Y.; Huang, D.; Xu, S.; Wang, Y.; Hu, N.; Yang, Z.; Xu, L. Microwave formation and photoluminescence mechanisms of multi-states nitrogen doped carbon dots. *Appl. Surf. Sci.* **2017**, *422*, 257–265. [CrossRef]
25. Farshbaf, M.; Davaran, S.; Rahimi, F.; Annabi, N.; Salehi, R.; Akbarzadeh, A. Carbon quantum dots: Recent progresses on synthesis, surface modification and applications. *Artif. Cells Nanomed. Biotechnol.* **2018**, *46*, 1331–1348. [CrossRef] [PubMed]
26. Boonmee, C.; Noipa, T.; Tuntulani, T.; Ngeontae, W.; Boonme, C. Cysteamine capped CdS quantum dots as a fluorescence sensor for the determination of copper ion exploiting fluorescence enhancement and long-wave spectral shifts. *Spectrochim. Acta Part A Mol. Biomol. Spectrosc.* **2016**, *169*, 161–168. [CrossRef]
27. Zhu, X.; Jin, H.; Gao, C.; Gui, R.; Wang, Z. Ratiometric, visual, dual-signal fluorescent sensing and imaging of pH/copper ions in real samples based on carbon dots-fluorescein isothiocyanate composites. *Talanta* **2017**, *162*, 65–71. [CrossRef]

28. Zhao, L.; Li, H.; Xu, Y.; Liu, H.; Zhou, T.; Huang, N.; Li, Y.; Ding, L. Selective detection of copper ion in complex real samples based on nitrogen-doped carbon quantum dots. *Anal. Bioanal. Chem.* **2018**, *410*, 4301–4309. [CrossRef]
29. Li, J.; Li, P.; Bian, M.; Huoa, D.; Hou, C.; Qin, H.; Zhang, S.-Y.; Zhang, L. A Fluorescent detection method for copper ions based on a direct redox route and desk study of wax-printed paper-based probes. *Anal. Methods* **2018**, *10*, 1895–1901. [CrossRef]
30. Zhang, W.J.; Liu, S.G.; Han, L.; Luo, H.Q.; Li, N.B. A ratiometric fluorescent and colorimetric dual-signal sensing platform based on N-doped carbon dots for selective and sensitive detection of copper(II) and pyrophosphate ion. *Sens. Actuators B Chem.* **2019**, *283*, 215–221. [CrossRef]
31. Li, H.; Bai, H.; Lv, Q.; Wang, W.; Wang, Z.; Wei, H.; Zhang, Q. A new colorimetric sensor for visible detection of Cu(II) based on photoreductive ability of quantum dots. *Anal. Chim. Acta* **2018**, *1021*, 140–146. [CrossRef] [PubMed]
32. Wang, Y.; Wu, W.; Wu, M.; Sun, H.-D.; Xie, H.; Hu, C.; Wu, X.-Y.; Qiu, J. Yellow-visual fluorescent carbon quantum dots from petroleum coke for the efficient detection of Cu2+ ions. *New Carbon Mater.* **2015**, *30*, 550–559. [CrossRef]
33. Chowdhury, S.; Rooj, B.; Dutta, A.; Mandal, U. Review on Recent Advances in Metal Ions Sensing Using Different Fluorescent Probes. *J. Fluoresc.* **2018**, *28*, 999–1021. [CrossRef] [PubMed]
34. Dager, A.; Uchida, T.; Maekawa, T.; Tachibana, M. Synthesis and characterization of Mono-disperse Carbon Quantum Dots from Fennel Seeds: Photoluminescence analysis using Machine Learning. *Sci. Rep.* **2019**, *9*, 1–12. [CrossRef]
35. Kumar, P.; Bhatt, G.; Kaur, R.; Dua, S.; Kapoor, A. Synthesis and modulation of the optical properties of carbon quantum dots using microwave radiation. *Full Nanotub. Carbon Nanostructures* **2020**, *28*, 724–731. [CrossRef]
36. Dots, Q. Facile Synthesis of Surface-Modified Carbon. *Materials* **2020**, *13*, 3313.
37. Rooj, B.; Dutta, A.; Islam, S.; Mandal, U. Green Synthesized Carbon Quantum Dots from Polianthes tuberose L. Petals for Copper (II) and Iron (II) Detection. *J. Fluoresc.* **2018**, *28*, 1261–1267. [CrossRef]
38. Smith, A.M.; Nie, S. Semiconductor Nanocrystals: Structure, Properties, and Band Gap Engineering. *Acc. Chem. Res.* **2010**, *43*, 190–200. [CrossRef]
39. Mai, X.-D.; Chi, T.T.K.; Nguyen, T.-C.; Ta, V.-T. Scalable synthesis of highly photoluminescence carbon quantum dots. *Mater. Lett.* **2020**, *268*, 127595. [CrossRef]
40. Liu, X.; Gao, W.; Zhou, X.; Ma, Y. Pristine graphene quantum dots for detection of copper ions. *J. Mater. Res.* **2014**, *29*, 1401–1407. [CrossRef]
41. Jiang, S.; Lu, Z.; Su, T.; Feng, Y.; Zhou, C.; Hong, P.; Sun, S.; Li, C.; Lu; Su; et al. High Sensitivity Detection of Copper Ions in Oysters Based on the Fluorescence Property of Cadmium Selenide Quantum Dots. *Chemosens* **2019**, *7*, 47. [CrossRef]
42. Wang, F.; Gu, Z.; Lei, W.; Wang, W.; Xia, X.; Hao, Q. Graphene quantum dots as a fluorescent sensing platform for highly efficient detection of copper(II) ions. *Sens. Actuators B Chem.* **2014**, *190*, 516–522. [CrossRef]
43. Rong, M.-C.; Zhang, K.-X.; Wang, Y.; Chen, X. The synthesis of B, N-carbon dots by a combustion method and the application of fluorescence detection for Cu^{2+}. *Chin. Chem. Lett.* **2017**, *28*, 1119–1124. [CrossRef]
44. Han, T.; Yuan, Y.; Kang, H.; Zhang, Y.; Dong, L. Ultrafast, sensitive and visual sensing of copper ions by a dual-fluorescent film based on quantum dots. *J. Mater. Chem. C* **2019**, *7*, 14904–14912. [CrossRef]

Publisher's Note: MDPI stays neutral with regard to jurisdictional claims in published maps and institutional affiliations.

Article

A Comparison of "Bottom-Up" and "Top-Down" Approaches to the Synthesis of Pt/C Electrocatalysts

Alexandra Kuriganova [1,*], Nikita Faddeev [1], Mikhail Gorshenkov [2], Dmitri Kuznetsov [1], Igor Leontyev [3] and Nina Smirnova [1,2]

1 Technological Department, Platov South-Russian State Polytechnic University (NPI), 346428 Novocherkassk, Russia; nikita.faddeev@yandex.ru (N.F.); kuznetsovdm@mail.ru (D.K.); smirnova_nv@mail.com (N.S.)
2 Department of Physical Materials Science, National University of Science and Technology "MISiS", 119049 Moscow, Russia; mvgorshenkov@gmail.com
3 Physical Department, Southern Federal University, 344090 Rostov-on-Don, Russia; i.leontiev@rambler.ru
* Correspondence: kuriganova_@mail.ru; Tel.: +7-908-197-5187

Received: 6 May 2020; Accepted: 28 July 2020; Published: 6 August 2020

Abstract: Three 40 wt % Pt/C electrocatalysts prepared using two different approaches—the polyol process and electrochemical dispersion of platinum under pulse alternating current—and a commercial Pt/C catalyst (Johnson Matthey prod.) were examined via X-ray diffraction (XRD) and transmission electron microscopy (TEM). The stability characteristics of the Pt/C catalysts were studied via long-term cycling, revealing that, for all cycling modes, the best stability was achieved for the Pt/C catalyst with the largest platinum nanoparticle sizes, which was synthesized via electrochemical dispersion of platinum under pulse alternating current. Our results show that the mass and specific electrocatalytic activities of Pt/C catalysts toward ethanol electrooxidation are determined by the value of the electrochemically active Pt surface area in the catalysts.

Keywords: nanoparticles; platinum catalyst; synthesis method; polyol process; electrochemical dispersion; alternating current; ethanol electrooxidation; electrocatalysis; fuel cell

---

1. Introduction

A vast amount of global experience has so far been accumulated in the creation of electrocatalytic materials for solid polymer fuel elements. In particular, elucidating the effect of composition and type (platinum or platinum-free systems) of catalyst [1,2], size and content of electroactive particles [3], composition and structure of support [4,5], and parameters that determine the performance of the catalyst at the nanolevel, i.e., adsorption site structure [6] or the presence of metal nanoparticle defects [7], on the efficiency of materials involved in the electrocatalytic processes running in proton exchange membrane fuel cells (PEMFCs) is within the scope of many relevant works.

In addition to the long-term study of platinum-free electrocatalytic systems [1,8,9], attempts are still being made to reduce the platinum content in the materials via the creation of composites, where platinum is partly replaced with a base metal (alloys or core–shell structures) [10–12]. So, real PEMFCs still operate on pure Pt/C catalysts with a large amount (40 wt % or higher) of platinum.The above parameters that determine the performance of the electrocatalytic material can be varied by making relevant changes in the synthesis route. Currently, Pt/C electrocatalysts can be obtained using two main approaches. The first is the so-called "bottom-up" approach, where platinum nanoparticles are produced via the chemical reduction of platinum ions to metal Pt particles using impregnation and microemulsion methods [13]. The other is represented by the "top-down" ways, where the formation of platinum particles is achieved by the fragmentation of bulk platinum to nanosized particles. Among these methods are ball-milling and laser ablation [14]. In other words, all top-down methods today are based primarily on physical phenomena. The use of chemical (or electrochemical) top-down methods for the synthesis of metal nanoparticles and, especially, supported nanocatalysts is a rather rare phenomenon. Although the influence of the synthesis conditions of Pt-based catalysts on their electrocatalytic activity has already been investigated in some works [15,16], only impregnation and microemulsion approaches were taken into account for this goal; that is, two chemical bottom-up methods were compared. Thus, in the present work, the properties of Pt/C catalysts obtained by two methods, which are both chemical but belong to the two different groups of methods (a top-down approach and a bottom-up approach), are compared for the first time.

2. Materials and Methods

2.1. Chemicals

In the presented work, the following chemicals and materials were used: hexachloroplatinic acid hexahydrate, $H_2[PtCl_6]{\cdot}6H_2O$ (ACS reagent, ≥37.50% Pt basis, Merck KGa, Darmstadt, Germany); sodium borohydride, $NaBH_4$ (ReagentPlus®, 99%, Sigma-Aldrich, USA); sodium hydroxide, NaOH (BioXtra, ≥98%, Sigma-Aldrich, St. Louis, Mo., USA); ethylene glycol, $C_2H_6O_2$ (ReagentPlus®, ≥99%, Sigma-Aldrich, USA); sulfuric acid, H_2SO_4 (99.999%, Sigma-Aldrich, USA); ethanol, CH_3CH_2OH (95%, Sigma-Aldrich, USA); ammonium hydroxide solution, NH_4OH (ACS reagent, 28.0–30.0% NH_3 basis, Sigma-Aldrich); acetone, CH_3COCH_3 (ACS reagent, ≥99.5%, Sigma-Aldrich, USA); platinum foil (thickness 0.5 mm, 99.99% trace metals basis, Sigma-Aldrich); and carbon, Black Vulcan XC 72R (Fuel Cell Store ©, College Station, TX, USA). Nafion™ perfluorinated resin, aqueous dispersion (10 wt % in H_2O, eq. wt. 1000), was also used. All solutions were prepared fresh daily in deionized water (purified by a Milli-Q water system).

2.2. Synthesis of Pt/C via the Polyol Process

A carbon support with a weight of 0.2 g and $H_2[PtCl_6]{\cdot}6H_2O$ solution were added to an ethylene glycol/water (75 mL/30 mL) organic mixture. The carbon support concentration in the electrolyte solution was 2 g L^{-1}. The suspension of carbon support in NaOH solution was homogenized under ultrasound for 30 min. An aqueous ammonium solution was then added dropwise to achieve a solution pH of 11; the mixture was stirred in a magnetic stirrer for another 30 min. After that, 15 mL of a freshly prepared 0.5M$NaBH_4$ solution was introduced under constant stirring (200 rpm); the mixture was then stirred for 50 min. At the end of the synthesis, the suspension was filtered and washed repeatedly with acetone and distilled water to achieve a neutral pH value. The powdered electrocatalyst was dried at a temperature of 75 °C until constant weight was reached. The catalyst obtained by the polyol process was marked as sample "CH".

2.3. Synthesis of Pt/C via EDPAC

Synthesis of Pt/C via the electrochemical dispersion pulsed alternating current (EDPAC) technique was carried out as follows. Two electrodes made of Pt foil, each with a surface area of 6 cm^2, were placed into the electrolyzer with Vulcan XC-72 carbon support suspended in 2M NaOH aqueous solution. The carbon support concentration in the electrolyte solution was 2 g L^{-1}. The suspension of carbon support in the NaOH solution was stirred and cooled to 45–50 °C before synthesis. During the synthesis, a pulsed alternating current with a density of 1 A/cm^2 (frequency 50 Hz) was applied to the platinum electrodes to disperse them into metal nanoparticles, as described in a previous work [17]. The synthesis was carried out with constant stirring (200 rpm). The metal loading in the catalyst was controlled via the synthesis time. At the end of the synthesis, the suspension was filtered and rinsed with distilled water to achieve a neutral pH value. The electrocatalyst powder was then dried at a temperature of 75 °C until constant weight was accomplished. The catalyst obtained by the EDPAC method was marked as sample "ED".

The platinum content in the synthesized Pt/C catalysts determined by thermogravimetry analysis.

2.4. Physical Characterization

Thermogravimetric measurements were made on a Mettler Toledo TGA/DSC 1 in the range of 303–1073 K at a heating rate of 10 K/min under an air atmosphere.

The X-ray diffraction (XRD) data were collected on a Bragg–Brentano D8 Advance Bruker laboratory diffractometer (40 kV/30 mA Cu Kα radiation) equipped with a lynxEye XE detector. The diffractograms were recorded in the angular range 2θ = 20–90 angle at a scan rate of 0.18 angle/min, with an accumulation time of 5 s per spectrogram. A standard powdered LaB6 sample (NISTSRM660a) was used to find the instrumental resolution function. The structural and microstructural characteristics of the composites were determined using the Rietveld refinement technique with two sets of linear combinations of spherical harmonics for the Laue class m3m. The background was simulated with $2\theta^{-6}$ coefficients of a polynomial function.

The measurements via transmission electron microscopy were made by means of a JEOL JEM 1400 microscope at an accelerating voltage of 120 kV in the light field. The phase composition of powders was determined from the electron diffraction patterns that were indexed using the crystallographic database. Samples were prepared via dispersion in ethanol and exposed to ultrasound for 10 min. A droplet with suspended particles was applied onto copper meshes covered with a carbon film (Holey Carbon Grid). Once the alcohol had evaporated, the meshes were placed in the microscope for further analysis.

2.5. Electrochemical Measurements

All electrochemical measurements were made using a standard three-electrode cell. A working electrode was made following a technique described in a previous work [17]. A platinum wire was used as the counter electrode, and an Ag/AgCl electrode was the reference electrode, although all potentials in the study are given relative to a reversible hydrogen electrode (RHE). The voltages were recalculated relative to the RHE using the Nernst equation and preliminarily measuring the pH values of the solutions used in the experiments.

The electrochemically active surface area (ECSA) was evaluated via CO stripping. The COadsorption was performed in a pre-deaerated background electrolyte (0.5 M H_2SO_4) at E = 0.3 V. The ECSA was calculated with respect to the charge used in CO_{ads} oxidation, given the fact that its value to oxidize a monolayer of CO adsorbed on Pt is equal to 420 μC cm^{-2}.

The electrocatalytic activity of the Pt/C catalysts was determined via cycling voltammetry through ethanol electrochemical oxidation (0.5 M H_2SO_4 + 0.5 M EtOH). To characterize the specific activity, the current was correlated with the electrochemically active surface area (cm^2) of platinum. The operation stability of the Pt/C electrocatalysts was evaluated via multiple cycling in 0.5MH_2SO_4 in obedience to stress testing modes [18,19], and the relevant data are listed in Table 1.

Table 1. Experimental conditions for evaluation of the operation stability of the Pt/C catalysts during cycling. The electrochemically active surface area (ECSA) during cycling was calculated using the H_{UPD} method.

Parameter	Accelerated Ageing, Mode 1	Soft Ageing, Mode 2
Number of cycles	1500	5000
Electrolyte	0.5 M H_2SO_4	0.5 M H_2SO_4
Temperature (°C)	22–23	22–23
Potential range (V)	0.05–1.35	0.6–1.0
Potential sweep rate (mV s^{-1})	50	100
ECSA measurement	Every 100 cycles	Every 500 cycles

The microstructural and electrochemical characteristics of the ED and CH Pt/C catalysts were compared with those of a commercial 40 wt % Pt/C catalyst (Johnson Matthey) that was marked as sample "JM".

3. Results and Discussion

3.1. Physical Characterization of Pt/C Electrocatalysts

All the synthesized Pt/C composites exhibited the presence of XRD reflexes corresponding to an fcc lattice (*Fm3m*) of platinum, as observed in Figure 1a. The microstructural characteristics, as well as the unit cell parameters of the catalysts, were found via Rietveld refinement of the diffractograms (see Table 1). According to the results, we have the following:

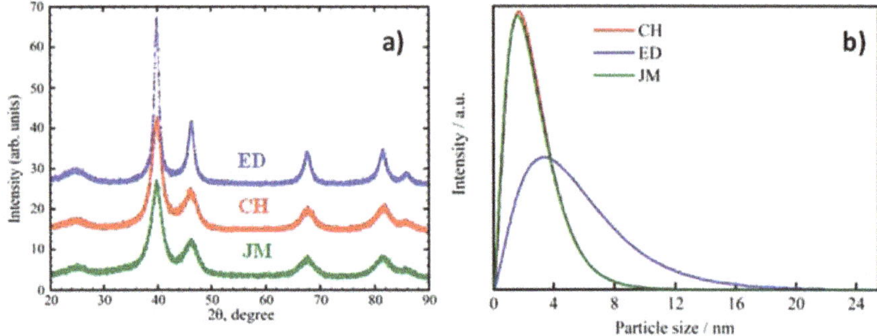

Figure 1. (**a**) XRD powder patterns of the Pt/C catalysts; (**b**) Particle size distribution of catalysts. The blue line refers to the catalyst prepared by the electrochemical dispersion pulsed alternating current technique (ED), the green curve is associated with commercial catalyst (JM), and the red curve is from the catalyst obtained by the polyol process (CH).

(i) The average D_{av} value of the Pt nanoparticles in the samples ranges from 2.5 to 4.7 nm; in the case of the ED catalyst prepared via the electrochemical method, the D_{av} value is almost twice as high as those of the commercial catalyst and the CH sample obtained via the polyol process.

(ii) The unit cell parameter of the Pt/C nanoparticles is smaller than that for bulk Pt, reducing with a decrease in Pt particle size on account of the size effect by analogy with a situation described in a previous work [20].

(iii) The anisotropy factor $R = D_{200}/D_{111}$, determining the Pt nanoparticle shape [21], is almost equal for all samples and refers to the R factor of a cuboctahedron.

(iv) The geometric surface area of the catalyst nanoparticles, calculated according to the technique reported in a previous work [22], as well as the average size and particle size distribution are given in Table 2. The particle size distributions are also shown in Figure 1b.

Table 2. XRD data of the Pt/C catalysts.

Sample	ED	CH	JM
D_{av} (nm)	4.74	2.55	2.48
ΔD_{av}	0.86	0.5	0.2
D_{111} (nm)	6.68	2.76	2.70
D_{200} (nm)	4.71	2.08	2.01
D_{200}/D_{111}	0.78	0.77	0.83
$<D>$ (nm)	5.56	2.69	2.68
σ (nm)	3.49	1.63	1.69
S_{geom} ($m^2\ g^{-1}$)	28.13	59.96	58.14
a (Å)	3.9153	3.9133	3.9151

It can be clearly seen in the TEM images (Figure 2a–c) of the catalysts that the most uniform distribution of platinum nanoparticles over the carbon support surface was observed in the JM sample. For the CH and JM composites, the Pt nanoparticle sizes were found by TEM to be 4.9 nm and 3.5 nm, respectively, while the ED sample demonstrated an average Pt nanoparticle size of about 10.4 nm. This may be due to a greater degree of agglomeration of platinum nanoparticles in the ED sample.

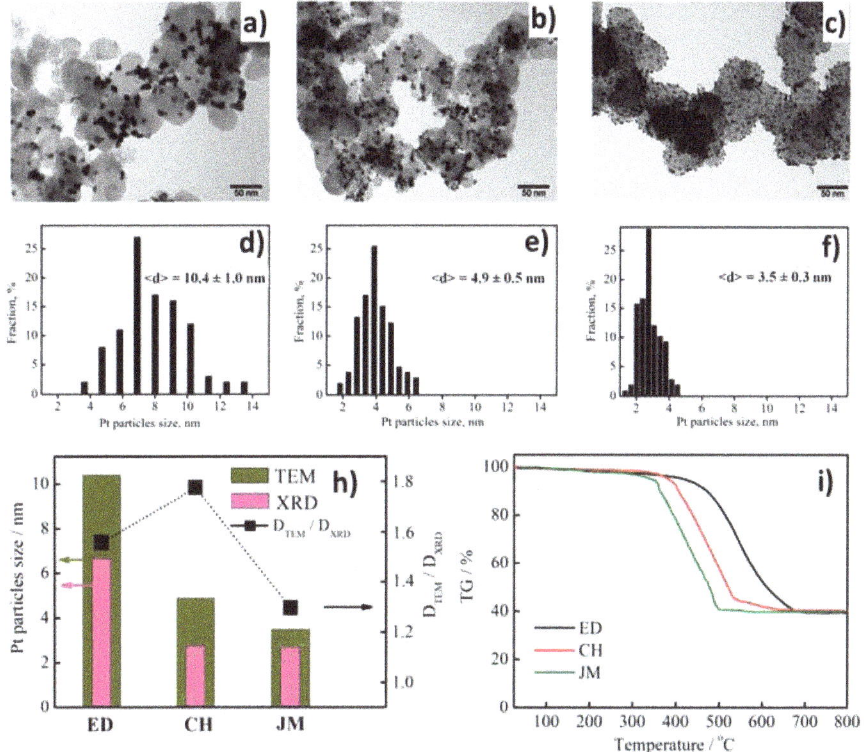

Figure 2. (**a**)–(**c**) Transmission electron microscopy images and (**d**)–(**f**) platinum particle size distributions in the Pt/C electrocatalysts: ED (**a**,**d**), CH (**b**,**e**), and JM (**c**,**f**) samples; a comparison of Pt particle sizes (D_{111}) evaluated via XRD and TEM (**h**); TGA of the ED, CH, and JM samples (**i**).

This tendency of Pt nanoparticle size variation depending on the catalyst synthesis conditions coincides with the XRD results (Figure 2h). Meanwhile, the ratio of Pt nanoparticle sizes evaluated via TEM to those provided by XRD (D_{TEM}/D_{XRD}) is much higher for the ED and CH composites than for the JM sample. This is testimony to the higher degree of agglomeration of ED and CH as compared to the commercial JM composite. That leads us to the following questions: What is the reason for such a large difference in the degree of platinum agglomeration in Pt/C samples obtained via different ways? Does it mean that agglomeration deteriorates the properties of the electrocatalyst?

Recall that the ED sample was produced via the "top-down" route, i.e., by the dispersion of platinum electrodes in the electrochemical cell at high cathode potentials. To interpret the mechanism of formation of platinum nanoparticles from bulk metal in the electrochemical system, we can offer the following explanation: platinum dispersion through (i) the emergence of intermetallic compounds of platinum with the electrolyte cation that is then decomposed by water (electrochemical dispersion) [17]; (ii) the formation of platinum anions stabilized with electrolyte cations that are precursors of platinum nanoparticles (cathodic corrosion) [23].

According to both approaches, platinum particles may form at only the cathode potentials. However, the dispersion of platinum in previous works [17,24,25] was implemented using pulsed electrolysis modes with short-term achievement of high anode potentials (>2 V) that make it possible to form three forms of chemisorbed oxygen and phase platinum oxides, the presence of which may simplify the intercalation of alkaline metal ions in the adsorption layer and thus increase the reaction rate. At low platinum contents in Pt/Al_2O_3 catalysts, X-ray absorption spectroscopy (XAS) revealed partly oxidized platinum nanoparticles PtO_x [24] that were not detected in a Pt/C (20% Pt) catalyst [17,25]. The important role of the anode current component, as well as that of the pH level of the near-electrode layer in terms of anodic dilution of noble metals during electrochemical release of oxygen was mentioned in several works published by Karl Mayrhofer's scientific group [26–31]. The ability to form colloidal Pt particles due to the formation of poorly soluble complexes [$Pt^{\delta+} \ldots O_2{}^{\delta-}$] on the electrode surface during charge transfer involving molecular oxygen was also noted in [32].

Thus, in spite of differences in understanding the mechanisms of formation of platinum-group metal nanoparticles via dispersion of a metallic electrode by electric current, it appears that the most important feature of the EDPAC method of producing Pt/C catalysts is the ability to form platinum nanoparticles in the near-electrode layer independently of the properties of the support. That, however, favors the presence of agglomerated platinum particles in the ED sample (Figure 2h) and broadens the platinum particle size distribution (Figures 1a and 2d).

Polyol synthesis of Pt/C catalysts using $NaBH_4$ as a reducing agent is one of the most common methods used to produce platinum-containing catalysts. The high reactivity of $NaBH_4$ as a reducing agent is among its most important benefits, allowing platinum to be reduced [33]. The use of ethylene glycol during the synthesis is due to the ability of glycolic acid anions to adsorb on the platinum particles' surface and to play the role of a stabilizing agent [34]. Regardless of the presence of a stabilizing agent, the high rate of Pt reduction impedes efficient control of the phase formation processes, resulting in nonuniform Pt nanoparticle distribution within a catalyst. Another reason for platinum agglomeration during the polyol synthesis of Pt/C catalysts is the influence of the carbon support surface state. As is known, the nature and state of the support exert a strong influence on the lattice strain, size, and distribution of mono- and polymetallic platinum-group metal nanoparticles [35–37].

Thus, in spite of the fact that both syntheses were carried out under identical hydrodynamic conditions, the degrees of agglomeration of Pt nanoparticles in the catalysts are different, which is due to the fundamentally different mechanisms of Pt nanoparticle formation during the polyol process and EDPAC synthesis.

According to TEM data, the commercial JM catalyst exhibits a uniform distribution of Pt nanoparticles within the support surface and a narrow nanoparticle size distribution (Figure 2c,f,h). In addition, with a decrease in the size of Pt nanoparticles in the catalyst, the temperature of the onset of thermal oxidation of carbon also decreased during thermogravimetric analysis of the Pt/C sample (Figure 2i). It should be also noted that the platinum content in the synthesized Pt/C catalysts determined by thermogravimetry was 40 ± 1 wt %.

The presence of platinum agglomerates in the ED catalyst reduces the electrochemically active surface area (ECSA) of ED. Only a part of the platinum surface participates in redox processes, decreasing the mass activity of the catalyst. Meanwhile, studies [38,39] have revealed a positive effect of agglomerated Pt particles on the kinetics of COoxidation, due to the fact that OH and CO species participating in the reaction are adsorbed on different nanoparticles, resulting in inter-particle processes.

3.2. Electrochemical Measurements

The electrochemically active surface area (ECSA) is one of most important characteristics of Pt/C electrocatalysts. To evaluate the surface area of porous materials, there is a frequently used method based on physical adsorption of nitrogen, the so-called Brunauer–Emmett–Teller BET method [40]. However, for Pt/C catalysts, most important is the surface area of the platinum particles, on which an electrochemical reaction is possible [41].

Among the most popular electrochemical methods for evaluation of the ECSA in Pt/C electrocatalysts is hydrogen underpotential deposition (H_{UPD}) [42]. Typically, to determine the ECSA with respect to the charge involved in hydrogen desorption (via the H_{UPD} method), cyclic voltammograms are recorded in different potential ranges [43]. Figure 3a displays part of the cyclic voltammogram (CV) curves (sample CH) for the H_2 adsorption/desorption region, acquired in 0.5 M H_2SO_4 under nitrogen in potential ranges of 0.03–1.3 V and 0.05–1.3 V vs. RHE. Even a slight shift in the cyclic potential boundary by 20 mV to the cathode range alters the ECSA value (the shaded area in Figure 3a), i.e., this method brings some experimental errors, which leads to an increase in the ECSA for Pt/C electrocatalysts (Table 2). Moreover, the accuracy in determining the ECSA via the H_{UPD} method may decrease because of a spill-over effect based on the transition (surface diffusion) of poorly adsorbed hydrogen from the Pt surface to the carbon support surface, resulting in a smaller amount of electricity consumed for adsorption/desorption of H_2 and, consequently, in a lower ECSA value.

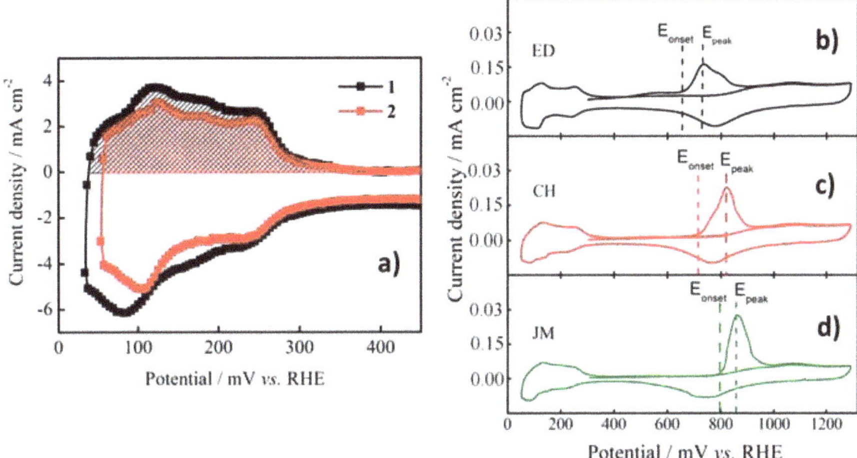

Figure 3. (**a**) H_2 adsorption/desorption range of cyclic voltammogram (CV) curves of Pt/C (sample CH) in 0.5 M H_2SO_4, N_2 atmosphere, potential ranges 0.03–1.3 V (1), 0.05–1.3 V vs. the reversible hydrogen electrode (RHE) (2); CO stripping on the Pt/C catalysts ED (**b**), CH (**c**), and JM (**d**) in electrolyte 0.5 M H_2SO_4 with a scan rate of 20 mV s^{-1}.

For this reason, the ECSAs of the Pt/C electrocatalysts were evaluated using CO stripping (Figure 3b,c). CO stripping is the most commonly discussed electrocatalytic reaction [44–48]. The mechanism of electrooxidation of COon a Pt surface is described by the Langmuir–Hinshelwood mechanism. The ECSA is determined with respect to the charge involved in CO_{ads} oxidation, given that the oxidation of a CO monolayer adsorbed on Pt requires a value of 420 μC cm^{-2}.

It was observed that the ECSA increased with decreasing Pt particle size in the catalyst at all degrees of particle agglomeration (Table 3, Figure 2h). The highest ECSA value was for the commercial JM catalyst at 26 ± 0.5 m^2 g^{-1}. It should be noted that in various studies (Table 3), the ECSA values of the JM catalyst (40 wt % Pt) differed from each other by more than a factor of 10 (from 18 to 190 m^2 g^{-1}), despite the fact that quite similar procedures were used to prepare the "catalytic inks" and working electrodes.

It is worth mentioning that the E_{onset} and E_{peak} values behave as a function of the average platinum nanoparticle size, being the highest for the JM Pt/C catalyst. This dependence seems to be due to a rise in the binding energy of COand the Pt surface with decreasing Pt nanoparticle size [48]. Among other reasons could be the difference in the degree of surface defectiveness of the particles that are the active centers for OH_{ads} adsorption and the variance in CO_{ads} surface mobility [45,48].

The most important functional feature of Pt/C electrocatalysts is their operation stability, maintaining stability of their ECSA. Degradation of a Pt/C electrocatalyst can be due to the dilution of platinum nanoparticles, their agglomeration and isolation by the ionomer, or oxidation of the carbon support [19,49,50]. The operation stability of the Pt/C electrocatalysts was evaluated using express tests in different voltammetric cyclic modes. The cyclic conditions corresponding to express testing modes were applied in this work [18,19] to study the operation stability of the electrocatalysts (Table 1).

During cycling of the Pt/C catalysts in Mode 1, the relative change in ECSA was 47% (ED) to 17.8% (CH). The ECSA of the ED sample decreased in a monotonic manner (Figure 4a), while the CH and JM samples exhibited high rates of degradation. The CHand JM composites lost more than 80% of their electrochemically active surface areas, with surface area losses of more than 40% at 500 cycles.

A similar tendency was observed during cycling in Mode 2. It is worth mentioning that the ECSA of the ED sample in this cycling mode reached a maximum value by only the 1000th cycle, which is due to the presence of partly oxidized Pt nanoparticles obtained via electrochemical dispersion [17,51] and, consequently, to slower catalytic activity upon cycling in a narrow potential range.

Table 3. Comparative ECSA values of Pt/C electrocatalysts obtained by polyol methods and the commercial JM catalyst in different labs.

Sample Preparation Method	Particle Size, nm	ECSA ($m^2\ g^{-1}$)/Technique	Ref.
JM (40 wt % Pt/C)	3.0	50.9/H_{UPD}	[52]
	3.5	46.4/H_{UPD}	[53]
	-	39.4/H_{UPD}	[54]
	3.1	49 ± 1/H_{UPD}	[55]
	4.6	60.8/H_{UPD}	[56]
	3.0	74.5/CO stripping	[57]
	3.5	29.0/H_{UPD}	[58]
	-	63.9/H_{UPD}	[59]
	3.09	43.9/H_{UPD}	[60]
	3.0	54.21/H_{UPD}	[61]
	3.0	46.4/H_{UPD}	[62]
	5.1	18.17/H_{UPD}	[63]
	3.4	36.3/H_{UPD}	[64]
	3.76	190.0/H_{UPD}	[65]
	3.5	18.0 ± 0.1/H_{UPD} [1]	This work
		22.0 ± 0.2/H_{UPD} [2]	This work
		26.0 ± 0.5/CO stripping [3]	This work
30 wt % Pt/C/polyol process	3.6	-	[66]
19 wt % Pt/C/polyol process	2.0	99.0 ± 10/H_{UPD}	[67]
40 wt % Pt/C/polyol process	2.9	-	[68]
40 wt % Pt/C/microwave-assisted polyol synthesis	2.6 ± 0.7	-	[69]
40 wt % Pt/C/polyol synthesis	3.1	58.6/H_{UPD}	[49]
40 wt % Pt/C/polyol synthesis	2.9	53.0/H_{UPD}	[34]
CH (40 wt% Pt/C)/polyol synthesis	2.55 (XRD)	13.0 ± 0.1/H_{UPD} [1]	This work
		15.0 ± 0.2/H_{UPD} [2]	This work
		22.0 ± 0.2/CO stripping [3]	This work

[1] Potential range of 0.05–1.30 V; [2] Potential range of 0.03–1.26 V; [3] E_{ads} = 0.3 V vs. RHE.

The electrocatalytic activity of the Pt/C catalysts was studied by using the example of ethanol oxidation reaction (EOR). Figure 5a,b shows the cyclic voltammograms (CVs) of the Pt/C catalysts in 0.5 M H_2SO_4 + 0.5 M EtOH electrolyte. The anodic CV range at E = 0.7–1.1 V exhibits the ethanol oxidation peak. The decrease in current density at potentials E ≥ 1.0 V is due to the removal of adsorbed ethanol intermediates and to the adsorption of oxygen-containing particles. The cathodic CV range also reveals the ethanol oxidation peak, but at potential values of E = 0.9 ÷ 0.4 V.

It is worth mentioning the influence of the ECSA of Pt/C catalysts on their mass (MA) and specific activity (SA) as determined using the current density of the ethanol oxidation peak in the anodic CV range. As seen in Figure 5c, the larger the ECSA, the lower the surface activity of the Pt/C catalysts, while the mass activity as a function of the ECSA follows the reverse trend.

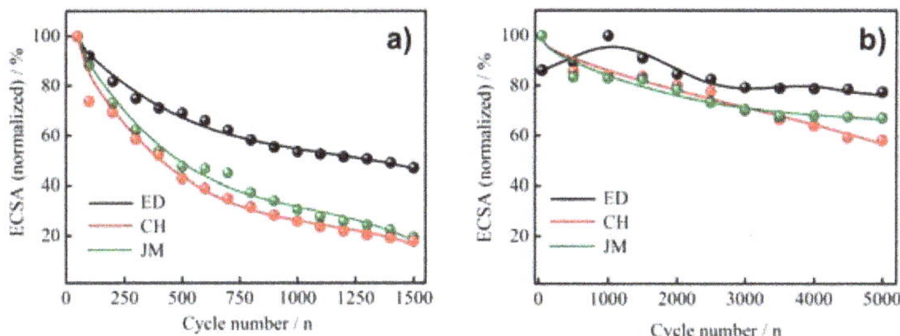

Figure 4. Stability of the Pt/C electrocatalysts as a function of the cyclic potential range: (**a**) 0.05–1.3 V vs. RHE; (**b**) 0.6–1.0 V vs. RHE.

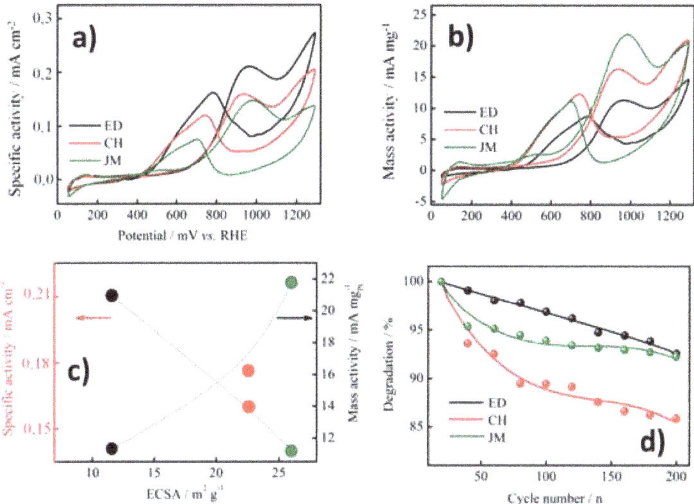

Figure 5. CV curves of the ED, CH, and JM Pt/C catalysts in 0.5 M H_2SO_4 + 0.5 M EtOH, scan rate 20 mV s^{-1} (**a**,**b**); specific activity and mass activity during the ethanol oxidation reaction (EOR) processes versus the ECSA of Pt/C catalysts, black balls—ED sample, red balls—CH sample, green balls—ED sample (**c**); durability of the ED, CH, and JM Pt/C catalysts in 0.5 M H_2SO_4 + 0.5 M EtOH (**d**).

The stability of the Pt/C catalysts (Figure 5d) obtained in 0.5 M H_2SO_4 electrolyte containing 0.5 M EtOH was estimated from the decrease in current density of the ethanol oxidation peak in the anodic CV range during 200 cycles of the Pt/C catalysts in the potential range of 0.05–1.3 V.

The ED composite exhibited a characteristic linear decrease in catalytic activity in the EOR. The commercial JM sample revealed a comparatively high degree of degradation in the first 50–100 cycles. However, by the 200th cycle, the degree of degradation of the JM and ED catalysts was ~7%, whereas that for the CH sample reached ~14% (Figure 5d). The relatively low tendency of the ED sample to degrade during cycling in both 0.5 M H_2SO_4 and 0.5 M H_2SO_4 + 0.5 M EtOH electrolyte seems to be thanks to the larger sizes of its Pt particles, which impede their quick agglomeration. Moreover, it was shown in a previous work [17] that Pt nanoparticles prepared via the electrochemical dispersion of platinum electrodes exhibit (100) crystallographic orientation of the Pt nanoparticles, which is more energetically favorable for the adsorption of oxygen atoms [70].

4. Conclusions

Two Pt/C catalysts were produced via fundamentally different methods. The ED sample obtained via electrochemical dispersion of platinum under the action of a pulsed alternating current (top-down) was characterized by a broad platinum nanoparticle size distribution; the average Pt nanoparticle size provided by TEM was 10.4 nm. The CH composite produced via the polyol (bottom-up) method revealed a narrow Pt nanoparticle size distribution in comparison to the ED sample; the average Pt nanoparticle size was found to be 4.9 nm. Both catalysts were compared with a commercial Pt/C composite (Johnson Matthey, or JM) that evidenced the lowest degree of platinum nanoparticle agglomeration (against ED and CH). The ECSA value of the commercial JM sample was shown to strongly differ in various literature sources. Thus, both the ECSA and the catalytic activity of Pt/C electrocatalysts depend on the working electrode formation method and the experimental conditions. Moreover, the mass and specific activities of the catalysts are first of all determined by the ECSA value. Thus, it is essential to compare the different Pt/C catalysts in terms of their mass and specific activities.

To conclude, the bottom-up approach enabled us to obtain a Pt/C electrocatalyst (CH sample) in which the average size of platinum nanoparticles was 4.9 nm and nanoparticles were uniformly distributed within a carbon support surface. On the one hand, a relatively high ECSA ensured high weight activity of the CH catalyst in the EOR process; on the other hand, the degree of degradation of the catalyst during long-term cycling in 0.5 M H_2SO_4 and 0.5 M H_2SO_4 + 0.5 M EtOH electrolytes was relatively high. The top-down method, namely, the electrochemical dispersion of platinum under the action of pulsed alternating current, allowed us to produce a Pt/C electrocatalyst (ED) with larger platinum nanoparticles (10.4 nm) and a comparatively low ECSA value that exhibited high stability during cycling, whether placed in the background electrolyte or in ethanol. Despite the fact that samples JM and CH have higher mass activity in comparison with ED, these samples also have a higher tendency to degrade. Based on this, it can be assumed that, as a result, the mass activity of samples JM and CH is at least that of the ED sample. However, to establish this as fact, it is necessary to carry out longer experiments not in model conditions but in a single cell of a fuel cell.

Author Contributions: Investigation, N.F., I.L., D.K., and M.G.; writing—original draft preparation, A.K.; supervision, N.S. All authors have read and agreed to the published version of the manuscript.

Funding: This work was supported by the Russian Foundation for Basic Research (grant no. 18-33-20064 mol_a_ved).

Conflicts of Interest: The authors declare no conflict of interest.

References

1. Lin, C.-Y.; Zhang, D.; Zhao, Z.; Xia, Z. Covalent Organic Framework Electrocatalysts for Clean Energy Conversion. *Adv. Mater.* **2017**, *30*, 1703646. [CrossRef]
2. Jung, N.; Chung, D.Y.; Ryu, J.; Yoo, S.J.; Sung, Y.-E. Pt-Based nanoarchitecture and catalyst design for fuel cell applications. *Nano Today* **2014**, *9*, 433–456. [CrossRef]
3. Antolini, E. Structural parameters of supported fuel cell catalysts: The effect of particle size, inter-particle distance and metal loading on catalytic activity and fuel cell performance. *Appl. Catal. B Environ.* **2016**, *181*, 298–313. [CrossRef]
4. Park, Y.-C.; Tokiwa, H.; Kakinuma, K.; Watanabe, M.; Uchida, M. Effects of carbon supports on Pt distribution, ionomer coverage and cathode performance for polymer electrolyte fuel cells. *J. Power Sour.* **2016**, *315*, 179–191. [CrossRef]
5. Sharma, S.; Pollet, B.G. Support materials for PEMFC and DMFC electrocatalysts—A review. *J. Power Sour.* **2012**, *208*, 96–119. [CrossRef]
6. Han, B.C.; Miranda, C.R.; Ceder, G. Effect of particle size and surface structure on adsorption of O and OH on platinum nanoparticles: A first-principles study. *Phys. Rev. B* **2008**, *77*, 075410. [CrossRef]
7. Chen, Q.-S.; Vidal-Iglesias, F.J.; Solla-Gullón, J.; Sun, S.-G.; Feliu, J.M. Role of surface defect sites: From Pt model surfaces to shape-controlled nanoparticles. *Chem. Sci.* **2012**, *3*, 136–147. [CrossRef]
8. Chen, Z.; Higgins, D.; Yu, A.; Zhang, L.; Zhang, J. A review on non-precious metal electrocatalysts for PEM fuel cells. *Energy Environ. Sci.* **2011**, *4*, 3167–3192. [CrossRef]
9. Brouzgou, A.; Song, S.; Tsiakaras, P. Low and non-platinum electrocatalysts for PEMFCs: Current status, challenges and prospects. *Appl. Catal. B Environ.* **2012**, *127*, 371–388. [CrossRef]

10. Pryadchenko, V.V.; Srabionyan, V.V.; Kurzin, A.A.; Bulat, N.V.; Shemet, D.B.; Avakyan, L.A.; Belenov, S.V.; Volochaev, V.A.; Zizak, I.; Guterman, V.E.; et al. Bimetallic PtCu core-shell nanoparticles in PtCu/C electrocatalysts: Structural and electrochemical characterization. *Appl. Catal. A Gen.* **2016**, *525*, 226–236. [CrossRef]
11. Yang, H. Platinum-Based Electrocatalysts with Core-Shell Nanostructures. *Angew. Chem. Int. Ed.* **2011**, *50*, 2674–2676. [CrossRef] [PubMed]
12. Zou, J.; Wu, M.; Ning, S.; Huang, L.; Kanga, X.; Chen, S. Ru@Pt Core-Shell Nanoparticles: Impact of the Atomic Ordering of the Ru Metal Core on the Electrocatalytic Activity of the Pt Shell. *ACS Sustain. Chem. Eng.* **2019**, *7*, 9007–9016. [CrossRef]
13. Sharma, G.; Kumar, D.; Kumar, A.; Al-Muhtaseb, A.H.; Pathania, D.; Naushad, M.; Mola, G.T. Revolution from monometallic to trimetallic nanoparticle composites, various synthesis methods and their applications: A review. *Mater. Sci. Eng. C* **2017**, *71*, 1216–1230. [CrossRef] [PubMed]
14. Jamkhande, P.G.; Ghule, N.W.; Bamer, A.H.; Kalaskar, M.G. Metal nanoparticles synthesis: An overview on methods of preparation, advantages and disadvantages, and applications. *J. Drug Deliv. Sci. Technol.* **2019**, *53*, 101174. [CrossRef]
15. Alegre, C.; Gálvez, M.E.; Moliner, R.; Baglio, V.; Aricò, A.; Lázaro, M.J.; Gresa, C.A. Towards an optimal synthesis route for the preparation of highly mesoporous carbon xerogel-supported Pt catalysts for the oxygen reduction reaction. *Appl. Catal. B Environ.* **2014**, *147*, 947–957. [CrossRef]
16. Hernández-Fernández, P.; Rojas, S.; Ocón, P.; Gómez de la Fuente, J.L.; San Fabian, J.; Sanza, J.; Peña, M.A.; Garcia-Garcia, F.J.; Terreros, P.; Fierro, J.L.G. Influence of the Preparation Route of Bimetallic Pt-Au Nanoparticle Electrocatalysts for the Oxygen Reduction Reaction. *J. Phys. Chem. C* **2007**, *111*, 2913–2923. [CrossRef]
17. Leontyev, I.; Kuriganova, A.; Kudryavtsev, Y.; Dkhil, B.; Smirnova, N. New life of a forgotten method: Electrochemical route toward highly efficient Pt/C catalysts for low-temperature fuel cells. *Appl. Catal. A Gen.* **2012**, *431*, 120–125. [CrossRef]
18. Shinohara, K.; Ohma, A.; Iiyama, A.; Yoshida, T.; Daimaru, A. Membrane and Catalyst Performance Targets for Automotive Fuel Cells by FCCJ Membrane, Catalyst, MEA WG. *ECS Meet. Abstr.* **2011**, *41*, 775–784. [CrossRef]
19. Zhang, S.; Yuan, X.-Z.; Hin, J.N.C.; Wang, H.; Friedrich, K.A.; Schulze, M. A review of platinum-based catalyst layer degradation in proton exchange membrane fuel cells. *J. Power Sour.* **2009**, *194*, 588–600. [CrossRef]
20. Leontyev, I.N.; Kuriganova, A.B.; Leontyev, N.G.; Hennet, L.; Rakhmatullin, A.; Smirnova, N.V.; Dmitriev, V. Size dependence of the lattice parameters of carbon supported platinum nanoparticles: X-Ray diffraction analysis and theoretical considerations. *RSC Adv.* **2014**, *4*, 35959–35965. [CrossRef]
21. Leontyev, I.N.; Belenov, S.V.; Guterman, V.E.; Haghi-Ashtiani, P.; Shaganov, A.P.; Dkhil, B. Catalytic Activity of Carbon-Supported Pt Nanoelectrocatalysts. Why Reducing the Size of Pt Nanoparticles is Not Always Beneficial. *J. Phys. Chem. C* **2011**, *115*, 5429–5434. [CrossRef]
22. Leontyev, I.N.; Kuriganova, A.B.; Allix, M.; Rakhmatullin, A.; Timoshenko, P.E.; Maslova, O.A.; Mikheykin, A.S.; Smirnova, N.V. On the Evaluation of the Average Crystalline Size and Surface Area of Platinum Catalyst Nanoparticles. *Phys. Status Solidi B* **2018**, *255*. [CrossRef]
23. Yanson, A.I.; Rodriguez, P.; Garcia-Araez, N.; Mom, R.V.; Tichelaar, F.D.; Koper, M.T.M. Cathodic Corrosion: A Quick, Clean, and Versatile Method for the Synthesis of Metallic Nanoparticles. *Angew. Chem. Int. Ed.* **2011**, *50*, 6346–6350. [CrossRef] [PubMed]
24. Doronkin, D.E.; Kuriganova, A.B.; Leontyev, I.N.; Baier, S.; Lichtenberg, H.; Smirnova, N.V.; Grunwaldt, J.-D. Electrochemically Synthesized Pt/Al2O3 Oxidation Catalysts. *Catal. Lett.* **2015**, *146*, 452–463. [CrossRef]
25. Novikova, K.; Kuriganova, A.; Leontyev, I.; Gerasimova, E.; Maslova, O.; Rakhmatullin, A.; Smirnova, N.V.; Dobrovolsky, Y. Influence of Carbon Support on Catalytic Layer Performance of Proton Exchange Membrane Fuel Cells. *Electrocatalysis* **2017**, *9*, 22–30. [CrossRef]
26. Cherevko, S.; Topalov, A.A.; Zeradjanin, A.R.; Katsounaros, I.; Mayrhofer, K.J.J. Gold dissolution: Towards understanding of noble metal corrosion. *RSC Adv.* **2013**, *3*, 16516. [CrossRef]
27. Cherevko, S.; Zeradjanin, A.R.; Topalov, A.A.; Kulyk, N.; Katsounaros, I.; Mayrhofer, K.J.J. Dissolution of Noble Metals during Oxygen Evolution in Acidic Media. *ChemCatChem* **2014**, *6*, 2219–2223. [CrossRef]
28. Katsounaros, I.; Meier, J.C.; Klemm, S.O.; Topalov, A.A.; Biedermann, P.U.; Auinger, M.; Mayrhofer, K.J.J. The effective surface pH during reactions at the solid–liquid interface. *Electrochem. Commun.* **2011**, *13*, 634–637. [CrossRef]

29. Klemm, S.O.; Karschin, A.; Schuppert, A.K.; Topalov, A.A.; Mingers, A.M.; Katsounaros, I.; Mayrhofer, K.J.J. Time and potential resolved dissolution analysis of rhodium using a microelectrochemical flow cell coupled to an ICP-MS. *J. Electroanal. Chem.* **2012**, *677*, 50–55. [CrossRef]
30. Topalov, A.A.; Cherevko, S.; Zeradjanin, A.R.; Meier, J.C.; Katsounaros, I.; Mayrhofer, K.J.J. Towards a comprehensive understanding of platinum dissolution in acidic media. *Chem. Sci.* **2014**, *5*, 631–638. [CrossRef]
31. Topalov, A.A.; Katsounaros, I.; Auinger, M.; Cherevko, S.; Meier, J.C.; Klemm, S.O.; Mayrhofer, K. Dissolution of Platinum: Limits for the Deployment of Electrochemical Energy Conversion? *Angew. Chem. Int. Ed.* **2012**, *51*, 12613–12615. [CrossRef]
32. Kasatkin, V.E.; Tytik, D.L.; Revina, A.A.; Busev, S.A.; Abaturov, M.A.; Vysotskii, V.V.; Roldugin, V.I.; Kazanskii, L.P.; Kuz'Min, V.I.; Gadzaov, A.; et al. Electrochemical synthesis of iron and platinum nanoparticles in deionized water. *Prot. Met. Phys. Chem. Surf.* **2015**, *51*, 973–979. [CrossRef]
33. Chen, J.; Jiang, C.; Yang, X.; Feng, L.; Gallogly, E.B.; Wang, R. Studies on how to obtain the best catalytic activity of Pt/C catalyst by three reduction routes for methanol electro-oxidation. *Electrochem. Commun.* **2011**, *13*, 314–316. [CrossRef]
34. Qi, J.; Jiang, L.; Jing, M.; Tang, Q.; Sun, G. Preparation of Pt/C via a polyol process—Investigation on carbon support adding sequence. *Int. J. Hydrogen Energy* **2011**, *36*, 10490–10501. [CrossRef]
35. Comignani, V.; Sieben, J.M.; Sanchez, M.D.; Duarte, M.M.E. Influence of carbon support properties on the electrocatalytic activity of PtRuCu nanoparticles for methanol and ethanol oxidation. *Int. J. Hydrogen Energy* **2017**, *42*, 24785–24796. [CrossRef]
36. Fraga, M.A.; Jordão, E.; Mendes, M.J.; Freitas, M.M.A.; Faria, J.L.; Figueiredo, J.L. Properties of Carbon-Supported Platinum Catalysts: Role of Carbon Surface Sites. *J. Catal.* **2002**, *209*, 355–364. [CrossRef]
37. Gurrath, M.; Kuretzky, T.; Boehm, H.P.; Okhlopkova, L.B.; Lisitsyn, A.S.; Likholobov, V.A. Palladium catalysts on activated carbon supports. *Carbon* **2000**, *38*, 1241–1255. [CrossRef]
38. López-Cudero, A.; Solla-Gullón, J.; Herrero, E.; Aldaz, A.; Feliu, J.M. CO electrooxidation on carbon supported platinum nanoparticles: Effect of aggregation. *J. Electroanal. Chem.* **2010**, *644*, 117–126. [CrossRef]
39. Maillard, F.; Schreier, S.; Hanzlik, M.; Savinova, E.R.; Weinkauf, S.; Stimming, U. Influence of particle agglomeration on the catalytic activity of carbon-supported Pt nanoparticles in CO monolayer oxidation. *Phys. Chem. Chem. Phys.* **2005**, *7*, 385–393. [CrossRef]
40. Brunauer, S.; Emmett, P.H.; Teller, E. Adsorption of Gases in Multimolecular Layers. *J. Am. Chem. Soc.* **1938**, *60*, 309–319. [CrossRef]
41. Watt-Smith, M.J.; Friedrich, J.M.; Rigby, S.P.; Ralph, T.R.; Walsh, F.C. Determination of the electrochemically active surface area of Pt/C PEM fuel cell electrodes using different adsorbates. *J. Phys. D Appl. Phys.* **2008**, *41*, 174004. [CrossRef]
42. Li, W.; Lane, A.M. Resolving the HUPD and HOPD by DEMS to determine the ECSA of Pt electrodes in PEM fuel cells. *Electrochem. Commun.* **2011**, *13*, 913–916. [CrossRef]
43. Thomas, J.M. *Handbook of Heterogeneous Catalysis*, 2nd Completely Revised and Enlarged Edition; Ertl, G., Knözinger, H., Schüth, F., Weitkamp, J., Eds.; Wiley: Hoboken, NJ, USA, 2008; Volume 8.
44. Abe, K.; Uchida, H.; Inukai, J. Electro-Oxidation of CO Saturated in 0.1 M HClO4 on Basal and Stepped Pt Single-Crystal Electrodes at Room Temperature Accompanied by Surface Reconstruction. *Surfaces* **2019**, *2*, 315–325. [CrossRef]
45. Arenz, M.; Mayrhofer, K.J.J.; Stamenković, V.; Blizanac, B.B.; Tomoyuki, T.; Ross, P.N.; Marković, N.M. The Effect of the Particle Size on the Kinetics of CO Electrooxidation on High Surface Area Pt Catalysts. *J. Am. Chem. Soc.* **2005**, *127*, 6819–6829. [CrossRef]
46. Farias, M.J.S.; Busó-Rogero, C.; Vidal-Iglesias, F.J.; Solla-Gullón, J.; Camara, G.A.; Feliu, J.M. Mobility and Oxidation of Adsorbed CO on Shape-Controlled Pt Nanoparticles in Acidic Medium. *Langmuir* **2017**, *33*, 865–871. [CrossRef] [PubMed]
47. Liu, Y.; Duan, Z.; Henkelman, G. Computational design of CO-tolerant Pt3M anode electrocatalysts for proton-exchange membrane fuel cells. *Phys. Chem. Chem. Phys.* **2019**, *21*, 4046–4052. [CrossRef] [PubMed]
48. Maillard, F.; Savinova, E.R.; Stimming, U. CO monolayer oxidation on Pt nanoparticles: Further insights into the particle size effects. *J. Electroanal. Chem.* **2007**, *599*, 221–232. [CrossRef]
49. Lafforgue, C.; Zadick, A.; Dubau, L.; Maillard, F.; Chatenet, M. Selected Review of the Degradation of Pt and Pd-based Carbon-supported Electrocatalysts for Alkaline Fuel Cells: Towards Mechanisms of Degradation. *Fuel Cells* **2018**, *18*, 229–238. [CrossRef]

50. Weber, P.; Werheid, M.; Janssen, M.; Oezaslan, M. Fundamental Insights in Degradation Mechanisms of Pt/C Nanoparticles for the ORR. ECS. *Transactions* **2018**, *86*, 433–445.
51. Kuriganova, A.; Leontyeva, D.V.; Smirnova, N. On the mechanism of electrochemical dispersion of platinum under the action of alternating current. *Russ. Chem. Bull.* **2015**, *64*, 2769–2775. [CrossRef]
52. Zhang, C.; Hu, J.; Wang, X.; Zhang, X.; Toyoda, H.; Nagatsu, M.; Meng, Y. High performance of carbon nanowall supported Pt catalyst for methanol electro-oxidation. *Carbon* **2012**, *50*, 3731–3738. [CrossRef]
53. Wang, X.X.; Tan, Z.H.; Zeng, M.; Wang, J.N. Carbon nanocages: A new support material for Pt catalyst with remarkably high durability. *Sci. Rep.* **2014**, *4*, 4437. [CrossRef] [PubMed]
54. Zhang, Z.; Wang, Y.; Wang, X. Nanoporous bimetallic Pt-Au alloy nanocomposites with superior catalytic activity towards electro-oxidation of methanol and formic acid. *Nanoscale* **2011**, *3*, 1663–1674. [CrossRef]
55. Garsany, Y.; Singer, I.L.; Swider-Lyons, K.E. Impact of film drying procedures on RDE characterization of Pt/VC electrocatalysts. *J. Electroanal. Chem.* **2011**, *662*, 396–406. [CrossRef]
56. Basu, D.; Basu, S. Synthesis and characterization of Pt-Au/C catalyst for glucose electro-oxidation for the application in direct glucose fuel cell. *Int. J. Hydrogen Energy* **2011**, *36*, 14923–14929. [CrossRef]
57. Fu, T.; Fang, J.; Wang, C.; Zhao, J. Hollow porous nanoparticles with Pt skin on a Ag-Pt alloy structure as a highly active electrocatalyst for the oxygen reduction reaction. *J. Mater. Chem. A* **2016**, *4*, 8803–8811. [CrossRef]
58. Saleh, F.S.; Easton, E.B. Assessment of the ethanol oxidation activity and durability of Pt catalysts with or without a carbon support using Electrochemical Impedance Spectroscopy. *J. Power Sour.* **2014**, *246*, 392–401. [CrossRef]
59. Sohn, H.; Xiao, Q.; Seubsai, A.; Ye, Y.; Lee, J.; Han, H.; Park, S.; Chen, G.; Lu, Y. Thermally Robust Porous Bimetallic (NixPt1–x) Alloy Mesocrystals within Carbon Framework: High-Performance Catalysts for Oxygen Reduction and Hydrogenation Reactions. *ACS Appl. Mater. Interfaces* **2019**, *11*, 21435–21444. [CrossRef]
60. Jiang, Z.-J.; Jiang, Z.-J.; Meng, Y. High catalytic performance of Pt nanoparticles on plasma treated carbon nanotubes for electrooxidation of ethanol in a basic solution. *Appl. Surf. Sci.* **2011**, *257*, 2923–2928. [CrossRef]
61. Jung, J.H.; Park, H.J.; Kim, J.; Hur, S.H. Highly durable Pt/graphene oxide and Pt/C hybrid catalyst for polymer electrolyte membrane fuel cell. *J. Power Sour.* **2014**, *248*, 1156–1162. [CrossRef]
62. Zeng, M.; Wang, X.X.; Tan, Z.H.; Huang, X.X.; Wang, J.N. Remarkable durability of Pt-Ir alloy catalysts supported on graphitic carbon nanocages. *J. Power Sour.* **2014**, *264*, 272–281. [CrossRef]
63. Sakthivel, M.; Drillet, J.-F. An extensive study about influence of the carbon support morphology on Pt activity and stability for oxygen reduction reaction. *Appl. Catal. B Environ.* **2018**, *231*, 62–72. [CrossRef]
64. Song, J.; Li, G.; Qiao, J. Ultrafine porous carbon fiber and its supported platinum catalyst for enhancing performance of proton exchange membrane fuel cells. *Electrochim. Acta* **2015**, *177*, 174–180. [CrossRef]
65. Naidoo, Q.-L.; Naidoo, S.; Petrik, L.; Nechaev, A.; Ndungu, P. The influence of carbon based supports and the role of synthesis procedures on the formation of platinum and platinum-ruthenium clusters and nanoparticles for the development of highly active fuel cell catalysts. *Int. J. Hydrogen Energy* **2012**, *37*, 9459–9469. [CrossRef]
66. Liu, Z.; Hong, L.; Tham, M.P.; Lim, T.H.; Jiang, H. Nanostructured Pt/C and Pd/C catalysts for direct formic acid fuel cells. *J. Power Sour.* **2006**, *161*, 831–835. [CrossRef]
67. Alekseenko, A.A.; Guterman, V.E.; Volochaev, V.A.; Belenov, S.V.; Vladimir, G. Effect of wet synthesis conditions on the microstructure and active surface area of Pt/C catalysts. *Inorg. Mater.* **2015**, *51*, 1258–1263. [CrossRef]
68. Zhou, Z.; Wang, S.; Zhou, W.; Wang, G.; Jiang, L.; Li, W.; Song, S.; Liu, J.; Sun, G.; Xin, Q. Novel synthesis of highly active Pt/C cathode electrocatalyst for direct methanol fuel cell. *Chem. Commun.* **2003**, 394–395. [CrossRef] [PubMed]
69. Lebègue, E.; Baranton, S.; Coutanceau, C. Polyol synthesis of nanosized Pt/C electrocatalysts assisted by pulse microwave activation. *J. Power Sour.* **2011**, *196*, 920–927. [CrossRef]
70. Gu, Z.; Balbuena, P.B. Absorption of Atomic Oxygen into Subsurfaces of Pt(100) and Pt(111): Density Functional Theory Study. *J. Phys. Chem. C* **2007**, *111*, 9877–9883. [CrossRef]

Article

PAC Synthesis and Comparison of Catalysts for Direct Ethanol Fuel Cells

Alexandra Kuriganova [1,*], Daria Chernysheva [1], Nikita Faddeev [1], Igor Leontyev [1,2], Nina Smirnova [1] and Yury Dobrovolskii [1,3]

1 Technological Department of Platov South-Russian State Polytechnic University (NPI), 346428 Novocherkassk, Russia; da.leontyva@mail.ru (D.C.); nikita.faddeev@yandex.ru (N.F.); i.leontiev@rambler.ru (I.L.); smirnova_nv@mail.com (N.S.); dobr@icp.ac.ru (Y.D.)
2 Department of Physics, Southern Federal University, 344090 Rostov-on-Don, Russia
3 Institute of Problems of Chemical Physics, Russian Academy of Sciences, 142432 Chernogolovka, Russia
* Correspondence: kuriganova_@mail.ru

Received: 19 May 2020; Accepted: 17 June 2020; Published: 20 June 2020

Abstract: Pt/C, $PtMO_n/C$ (M = Ni, Sn, Ti, and PtX/C (X = Rh, Ir) catalyst systems were prepared by using the pulse alternating current (PAC) technique. Physical and electrochemical parameters of samples were carried out by x-ray powder diffraction (XRD), transmission electron microscopy (TEM), and CO stripping. The catalytic activity of the synthesized samples for the ethanol electrooxidation reaction (EOR) was investigated. The XRD patterns of the samples showed the presence of diffraction peaks characteristic for Pt, NiO, SnO_2, TiO_2, Rh, and Ir. The TEM images indicate that the Pt, Rh, and PtIr (alloys) particles had a uniform distribution over the carbon surface in the Pt/C, PtRh/C, PtIr/C, and $PtMO_n/C$ (M = Ni, Sn, Ti) catalysts. The electrochemically active surface area of catalysts was determined by the CO-stripping method. The addition of a second element to Pt or the use of hybrid supported catalysts can evidently improve the EOR activity. A remarkable positive affecting shift of the onset potential for the EOR was observed as follows: $PtSnO_2/C$ > $PtTiO_2/C$ ≈ PtIr/C ≈ PtNiO/C > PtRh/C ≈ Pt/C. The addition of SnO_2 to Pt/C catalyst led to the decrease of the onset potential and to significantly facilitate the EOR. The long-term cyclic stability of the synthesized catalysts was investigated. Thereby, the $PtSnO_2/C$ catalyst prepared by the PAC technique can be considered as a promising anode catalyst for direct ethanol fuel cells.

Keywords: direct ethanol fuel cell; platinum-based catalyst; electrocatalysis; pulse alternating current

1. Introduction

Fuel cells (FCs) have attracted much attention due to limited energy resources and environmental factors worldwide [1]. Conversion of the chemical energy of the fuel into electrical energy occurs in one stage in FCs. The advantages of FCs are undeniable compared to traditional energy sources. These include zero emissions into the atmosphere, low working temperature, and high efficiency. It is customary to use hydrogen as fuel in a FC. Hydrogen has the largest energy reserve per unit weight. However, difficulties arise when storing hydrogen. The most effective form for transporting and storing this chemical element is a liquid state. However, gas can be converted to liquid form only at a temperature of −253 °C, which requires special containers, equipment, and considerable financial costs.

FC with direct oxidation of liquid fuel is an alternative to hydrogen FC. Liquid fuel (ethanol and methanol) have several advantages over hydrogen: ease of transportation, storage, use, and low cost [2]. Ethanol is a renewable biofuel derived from biomass by fermentation [3]. This is why in recent years, more and more attention has been paid to direct ethanol fuel cells (DEFCs). The performance of fuel cells using ethanol as fuel is not inferior to the performance of methanol fuel cells [4].

The efficiency of DEFCs is determined by parameters such as the properties of the polymer electrolyte membrane, gas diffusion layer, structural decisions, the appropriate temperature, and balance in the system. However, the most important factor is the activity of the catalyst. Platinum on carbon support (Pt/C) is commonly used as the catalyst in FC. It is well known that the surface of the catalyst is a critical factor influencing its performance. Therefore, high surface area materials have been developed. Typically, the active phase is dispersed on a conductive support such as carbon. However, pure platinum is not the most efficient anodic catalyst for DEFCs [5].

The ethanol oxidation reaction (EOR) at the platinum consists of parallel reactions. For successful ethanol electrooxidation, the catalyst must adsorb the ethanol molecule, contributing to the dissociation of the C–C and C–H bonds. This will lead to the formation of adsorbed intermediates [6]. In addition, it must alleviate the reaction of the adsorbed intermediates with some oxygen containing particles to form carbon dioxide, formic, or acetic acids. Thus, a balance between the O-containing particles and adsorbed organic molecules should be maintained on the surface of the catalyst. This problem can be solved by increasing the electrolyte pH. This will lead to a change of EOR kinetics [7]. However, the most effective is to use a two-component catalyst including the alloys of platinum with oxygen adsorbing metals (Ir [8], Rh [9]) or using metal oxides (NiO [10], SnO_2 [11], TiO_2 [12]) as a catalyst support. In this case, the ethanol molecules adsorb on the platinum surface, while the oxygen adsorbing metal or metal oxides provide the presence of O-containing particles in the reaction zone.

In this study, we used the pulse alternating current (PAC) technique for the synthesis of Pt/C, $PtMO_n$/C (M = Ni, Sn, Ti), and PtX/C (X = Rh, Ir) catalyst systems and compared their electrocatalytic properties in the ethanol oxidation reaction (EOR).

2. Materials and Methods

The synthesis of Pt, NiO, SnO_2, TiO_2, Rh, and PtIr nanoparticles occurred under the pulse alternating current (duty cycle 25%). The ratio of the current density between the anodic and cathodic pulses was 1:1. The $PtMO_n$/C (M = Ni, Sn, Ti) and the PtRh/C catalysts were prepared in two stages. In the first stage, two metal (Ni, Sn, Ti, Rh) electrodes (thickness of 0.3 mm) were immersed in an electrolyzer with a Vulcan XC-72 carbon support suspended in 2 M NaOH (for Ti and Ni) or a 2 M NaCl (for Sn and Rh) aqueous solution. The suspension was permanently stirred and cooled to 45–50 °C. The metal electrodes were connected to an alternating current source. The formation of MO_n particles and Rh nanoparticles occurred under the influence of PAC on the electrodes. The as-prepared MO_n/C or Rh/C powders were collected by filtration and washed with distilled water. In the second stage, two Pt electrodes were immersed in the suspension of MO_n/C or Rh/C support in a 2 M NaOH aqueous solution. PAC was applied to the electrodes again and platinum nanoparticles were dispersed into the electrolyte. The prepared $PtMO_n$/C (M = Ni, Sn, Ti) and PtRh/C catalysts catalyst was rinsed with H_2O to a neutral pH. Catalysts were dried at 80 °C until a constant weight. Pt/C and PtIr/C catalysts were obtained using Pt and PtIr (alloy) electrodes in a carbon (Vulcan XC-72) suspension in 2 M NaOH in one stage. The metal phase loading in the catalyst or metal oxide loading in the catalyst support was determined by the weight loss of the metal electrodes (Pt, Ni, Sn, Ti, Rh, PtIr) and the results of thermogravimetry. The results of thermogravimetry showed a correlation with the theoretical composition of the catalysts, calculated from the weight loss of the metal electrodes (Pt, Ni, Sn, Ti, Rh, and PtIr) during synthesis.

The x-ray powder diffraction (XRD) studies were performed with an ARL X'TRA powder diffractometer, Thermo Scientific, (Cu K_α target, λ = 1.5418 Å). XRD data were collected in the 2θ range of 20°–85° using step scan mode. The average grain size was determined using the Scherrer equation:

$$D = K\lambda/(\text{FWHM}\cdot\cos\theta) \quad (1)$$

where D is the volume averaged grain size; λ is the x-ray wavelength; FWHM is the full width at half maximum of the diffraction peak; K = 0.89 is the dimensionless particle shape coefficient (Scherrer constant); and θ is a diffraction angle. The unit cell parameters (a) were determined with the UnitCell program using all peaks in the X-ray diffraction pattern.

Transmission electron microscopy (TEM) images were obtained with a JEM-2100 unit (200 kV). Suspension of a catalyst sample in hexane was prepared to study the obtained catalysts by TEM. The sample for analysis was prepared by depositing a droplet of the suspension onto an amorphous carbon-coated copper grid.

All electrochemical measurements were performed at room temperature using a standard three-electrode electrochemical cell and a P-45-X potentiostat (Elins, Russia). Electrochemical properties of the prepared catalyst systems were investigated by cyclic voltammetry (CV). A glassy carbon plate served as a working electrode, which was coated with "catalytic ink" prepared using 10 wt% Nafion dispersion, according to the technique [13]. A Pt wire was an auxiliary electrode, and an Ag/AgCl electrode was a reference electrode. All potentials presented in this work are quoted versus a reversible hydrogen electrode (RHE). The electrochemically active surface area (ECSA) was evaluated via CO-stripping. The CO adsorption was performed in a pre-deaerated background electrolyte (0.5 M H_2SO_4) at E = 0.3 V vs. RHE. The ECSA was calculated from the charge value used for CO_{ads} oxidation, taking into account that the oxidation of a carbon monoxide monolayer on Pt needs 420 $\mu C \cdot cm^{-2}$ [14]. The electrocatalytic properties were investigated in the 0.5 M C_2H_5OH + 0.5 M H_2SO_4 solution over a potential range between 0.05 and 1.3 V (RHE).

The long-term cyclic stability of the synthesized Pt/C, $PtMO_n/C$ (M = Ni, Sn, Ti) and PtX/C (X = Rh, Ir) catalysts was investigated by cycling in 0.5 M H_2SO_4 over a potential range of 0.05–1.3 V vs. RHE.

3. Results and Discussion

The metal or metal oxide content in the catalyst system was controlled by the time of synthesis based on the preliminarily found dispersion rates for the Pt, Ni, Sn, Ti, Rh, and PtIr electrodes. The dispersion rates were determined experimentally and amounted to 21, 38, 45, 50, 12, and 19 $mg \cdot cm^{-2} \cdot h^{-1}$ for Pt, Ni, Sn, Ti, Rh, and PtIr, respectively. The compositions of the obtained catalysts are presented in Table 1.

Table 1. Characteristics of Pt/C, PtNiO/C, $PtSnO_2/C$, $PtTiO_2/C$, PtRh/C, and PtIr/C catalysts.

Sample	Composition,%	*a*, Å	D, nm (Pt Nps)		ECSA, $m^2 \cdot g^{-1}$ (CO Stripping)	E_{pic}, V (RHE)	E_{onset}, V (RHE)
			XRD	TEM		CO-Stripping	EOR
Pt/C	Pt-25 C-75	Pt-3.9178	9.0 ± 1.0	8.0 ± 0.8	13.2 ± 1.0	0.585	0.675
PtNiO/C	Pt-25 NiO-30 C-45	Pt-3.9178 NiO-4.194	9.3 ± 1.0	6.0 ± 0.6	16.0 ± 1.0	0.616	0.593
$PtSnO_2/C$	Pt-25 SnO_2-30 C-45	Pt-3.9178 SnO_2-4.739	14.8 ± 1.0	6.0 ± 0.7	18.1 ± 1.0	0.442	0.535
$PtTiO_2/C$	Pt-25 TiO_2-30 C-45	Pt-3.9178 TiO_2-3.796	10.4 ± 1.0	6.0 ± 0.7	9.0 ± 1.0	0.475	0.559
PtRh/C	Pt-17 Rh-8 C-75	Pt-3.9178 Rh-3.8022	8.2 ± 1.0	7.0 ± 0.6	14.4 ± 1.0	0.656	0.575
PtIt/C	$Pt_{90}Ir_{10}$-25 C-75	PtIr-3.9149	5.5 ± 1.0	5.0 ± 0.5	12.3 ± 1.0	0.507	0.569

The prepared materials were investigated by XRD. The XRD patterns of the Pt/C, PtRh/C, PtIr/C, and Pt/MO_n/C catalysts synthesized via the PAC-technique are shown in Figure 1.

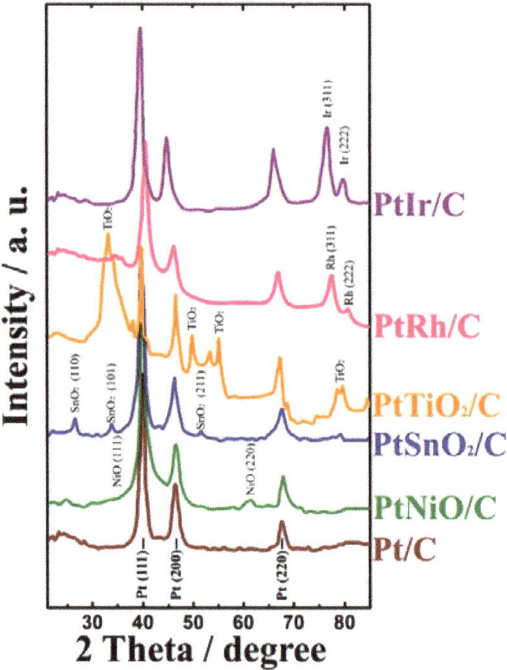

Figure 1. X-ray powder diffraction patterns of the Pt/C, PtNiO/C, PtSnO_2/C, PtTiO_2/C, PtRh/C, and PtIr/C catalysts.

XRD analysis showed the presence of a face-centered cubic structure Pt (Fm3m (No. 225)) in all of the catalyst samples. The analysis of the XRD patterns indicated the presence of titanium dioxide as anatase and rutile, tin dioxide with a tetragonal structure of SnO_2 (P42/mnm (No. 136)), and a nickel oxide cubic structure (Fm3m, JCPDS 75-0269) for the PtTiO_2/C, PtSnO_2/C, and PtNiO/C samples, respectively. Diffraction peaks characteristic for Rh and Ir were similarly observed in the PtRh/C and PtIr/C catalysts, respectively. The average size of crystallites in the synthesized catalysts was calculated from the data of the x-ray diffraction analysis (Figure 1, Table 1).

Figure 2 depicts the TEM image of the as-prepared catalysts. The surface average particle sizes presented in Table 1 were obtained by measuring the sizes of 200 particles randomly selected from the TEM images. Transmission electron microscopy studies of the Pt/C, PtRh/C, PtIr/C, and PtMO_n/C catalysts revealed that the Pt, Rh, and PtIr (alloys) particles had a uniform distribution over the carbon surface. In addition, the platinum nanoparticles had a uniform distribution over the surface of the metal oxides. However, increased Pt agglomeration was seen from the TEM images of the PtSnO_2/C (Figure 2c).

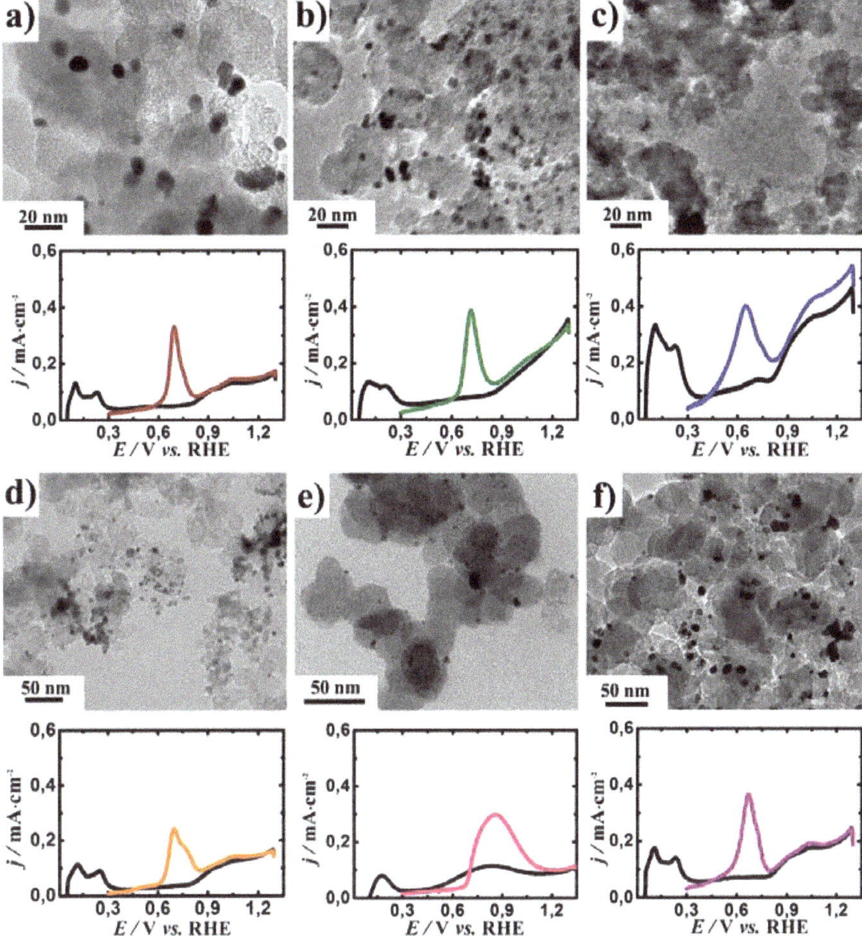

Figure 2. Transmission electron microscopy images and anodic scan of cyclic voltamperograms of (**a**) Pt/C; (**b**) PtNiO/C; (**c**) $PtSnO_2/C$; (**d**) $PtTiO_2/C$; (**e**) PtRh/C; and (**f**) PtIr/C catalysts in 0.5 M H_2SO_4 in the presence of adsorbed CO at E = 0.3 V (RHE), potential scan rate 20 $mV{\cdot}s^{-1}$.

The ECSA is one of the essential characteristics of electrocatalysts. The surface area of porous materials is typically estimated by a method based on physical nitrogen adsorption [15]. It is worth noting that the method is not suitable for the determination of the ECSA of platinum electrocatalysts, because in addition to the surface area of Pt nanoparticles undergoing the electrochemical process, it also considers the area of a carbon support that serves as an electron conductor in the catalyst. The ECSA of Pt-containing catalysts can usually be evaluated by means of electrochemical approaches [16] that take into account the real conditions of electrochemical processes. CV was used to obtain the ECSA of the catalysts. Figure 2 shows the anodic scan of the CV curve of the obtained catalysts in the 0.5 M H_2SO_4 electrolyte with a CO adsorbed ad-layer. The calculated values for the ECSA of the studied catalysts were about 14 ± 5 m^2/g_{Pt} (Table 1).

The important parameters for the DEFC anode catalysts are the overpotential for CO_{ad} and ethanol oxidation. The Pt/C electrocatalyst showed the onset potential for the CO oxidation of 0.585 V. For all the other studied catalytic systems, except for PtRh/C and PtNiO/C, the presence of the second component resulted in reducing the overpotential for CO oxidation. For $PtTiO_2/C$ and $PtSnO_2/C$, the CO oxidation peaks shifted by 0.110 V and 0.143 V to the cathode region, respectively (Table 1, Figure 2).

The as-prepared catalysts were also tested for the EOR (Figure 3a,b). The CV curves of the platinum nanoparticles supported on carbon (Pt/C) and hybrid metal oxide–carbon (PtNiO/C, $PtSnO_2/C$, and $PtTiO_2/C$) as well as bimetallic catalysts (PtRh/C and PtIr/C) were recorded in 0.5 M C_2H_5OH + 0.5 M H_2SO_4 solution. The presence of the metal oxide in the support or second metal in the bimetallic catalytic system caused a cathodic shift of the onset potential of the EOR (Table 1). The maximum effect was observed for the $PtSnO_2/C$ system where the onset potential of the EOR shifted toward negative potential values by 0.158 V. The positive effect of metal oxide in support of the catalyst for CO_{ad} and ethanol oxidation reaction can be explained by the theory of bifunctional catalysis [17]. In addition, electronic ligand effects are also possible [18]. Numerous studies have shown that the introduction of Sn in the composition of the catalyst for the oxidation of organic compounds has a positive effect on the process. A positive effect was observed due to earlier and easier adsorption of OH_{ads} on the surface of SnO_2 [19,20]. Rhodium displayed a higher current in the potential region of 0.4–0.6 V compared to the other catalysts, which can be due to the adsorption of O-containing species on the Rh surface (Figure 2f) and a little part is related to the oxidation of alcohol. The cathodic shift of the onset potential for EOR on the PtIr/C catalyst may be associated with the modification of the electronic structure of the PtIr alloy compared to Pt [21]. As a result, the adsorption energy of O-containing particles required for the complete oxidation of ethanol chemisorption products decreased.

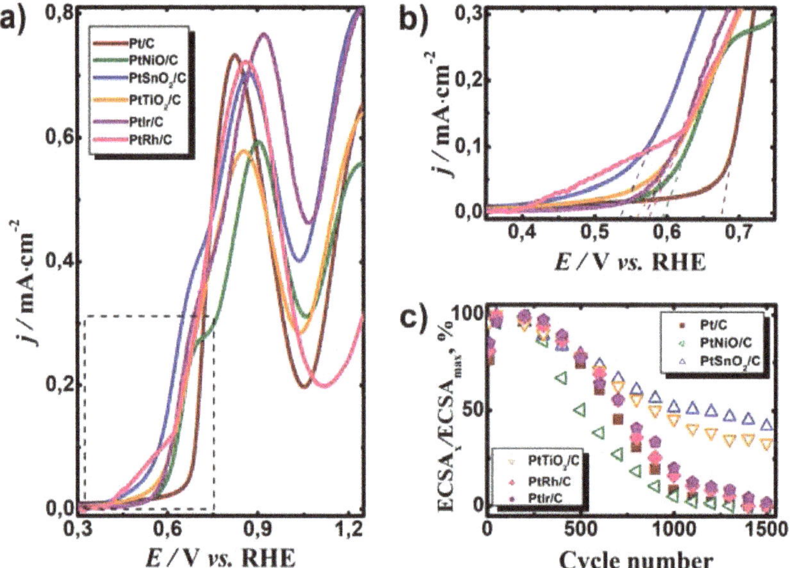

Figure 3. (**a**) The forward scans of cyclic voltamperograms of catalysts in 0.5 M C_2H_5OH + 0.5 M H_2SO_4 solution; potential scan rate was 5 mV·s^{-1}, (**b**) enlarged forward scans of cyclic voltamperograms, (**c**) the dependence electrochemically active surface area of catalysts, which was normalized to the maximum electrochemically active surface area of catalysts from the cycle numbers.

The stability of the synthesized catalysts was investigated by cycling in 0.5 M H_2SO_4 over a potential range of 0.05–1.3 V vs. RHE. Figure 3c shows the dependence ECSA of the catalysts, which was normalized to the maximum ECSA of catalysts from the cycle numbers. For the first 100 cycles, the value of the ECSA increased due to the purification of the platinum surface. However, after the 300th cycle, a decrease in the ECSA of all samples was observed. After a long-term cyclic stability test, the largest decrease in ECSA was observed in the Pt/C, PtNiO/C, PtRh/C, and PtIr/C samples. The fact is that the stability of the electrocatalyst in the FC is largely dependent on the stability of the catalyst support. As electrocatalysts work under very hard conditions, in the presence of the active component of the catalyst, carbon support oxidation occurs, which causes a loss of active component contact with the carbon support and thus the active component is out of the reaction zone [22]. It is assumed that the introduction of a metal oxide in the support will have a positive impact on the electrocatalyst's durability. The catalyst based on the hybrid support containing SnO_2 or TiO_2 was significantly more stable. The value of the ECSA after 1500 cycles decreased to 58% and 40% for $PtSnO_2/C$ and $PtTiO_2/C$, respectively. The behavior of the PtNiO/C electrocatalyst was the same as the Pt/C catalyst, but the rate of catalyst degradation was somewhat higher than that of Pt/C, which is probably due to the solubility of nickel oxide in the acidic medium.

4. Conclusions

In conclusion, the Pt/C, PtMOn/C (M = Ni, Sn, Ti) and PtX/C (X = Rh, Ir) catalyst systems were successfully prepared by using the pulse alternating current (PAC) technique. The XRD patterns of the samples showed the presence of diffraction peaks characteristic for Pt, NiO, SnO_2, TiO_2, Rh, and Ir. The TEM images indicate that the Pt, Rh, and PtIr (alloys) particles had a uniform distribution over the carbon surface in the Pt/C, PtRh/C, PtIr/C, and $PtMO_n/C$ (M = Ni, Sn, Ti) catalysts. When SnO_2/C and TiO_2/C were present in the supports of the catalyst, the onset potentials for carbon monoxide oxidation shifted in the cathodic direction by 0.143 V and 0.110 V, respectively, compared to the Pt/C catalyst.

All the obtained $PtMO_n/C$ and PtX/C catalyst systems revealed a better EOR performance compared to the conventional platinum catalyst. The addition of a second element to Pt or use of a hybrid support can evidently improve the EOR efficiency. A remarkable positive affecting shift of the onset potential for the EOR was observed as follows: $PtSnO_2/C > PtTiO_2/C \approx PtIr/C \approx PtNiO/C > PtRh/C \approx Pt/C$. The addition of SnO_2 to the Pt/C catalyst led to a decrease in the onset potential and significantly facilitated the EOR. In addition, the catalyst based on the hybrid support containing SnO_2 was significantly more stable than the Pt/C catalyst.

Thereby, the $PtSnO_2/C$ catalyst prepared by the PAC technique could be considered as a promising anode catalyst for DEFCs. In this way, this study contributed to the development of fuel cell technology with the direct oxidation of liquid fuels.

Author Contributions: Formal analysis, I.L.; Investigation, A.K., D.C., and I.L.; Supervision, N.S.; Visualization, N.F.; Writing—original text, N.F.; Writing—review & editing, Y.D. All authors have read and agree to the published version of the manuscript.

Funding: The article was prepared based on the results obtained during the implementation of the state assignment to carry out research on the project "Hydrogen fuel cells for small unmanned aerial vehicles: modeling, development, research" through the Ministry of Science and Higher Education of the Russian Federation (scientific code FENN-2020-0020).

Acknowledgments: This work was done using equipment at the Center for Shared Scientific Equipment "Nanotechnology" of the Platov South-Russian State Polytechnic University (NPI).

Conflicts of Interest: The authors declare no conflict of interest.

References

1. Li, Y.; Yang, J.; Song, J. Structure models and nano energy system design for proton exchange membrane fuel cells in electric energy vehicles. *Renew. Sust. Energy Rev.* **2017**, *67*, 160–172. [CrossRef]
2. Ghosh, S.; Maiyalagan, T.; Basu, R.N. Recent Advances in Nanostructured Electrocatalysts for Low-temperature Direct Alcohol Fuel Cells. *Electrocatal. Low Temp. Fuel Cells Fundam. Recent Trends* **2017**, *324*, 347.
3. Perez, J.; Paganin, V.A.; Antolini, E. Particle size effect for ethanol electro-oxidation on Pt/C catalysts in half-cell and in a single direct ethanol fuel cell. *J. Electroanal. Chem.* **2011**, *654*, 108–115. [CrossRef]
4. Wang, J.; Wasmus, S.; Savinell, R.F. Evaluation of Ethanol, 1-Propanol, and 2-Propanol in a Direct Oxidation Polymer-Electrolyte Fuel Cell A Real-Time Mass Spectrometry Study. *J. Electrochem. Soc.* **1995**, *142*, 4218–4224. [CrossRef]
5. Antolini, E. Catalysts for direct ethanol fuel cells. *J. Power Sources* **2007**, *170*, 1–12. [CrossRef]
6. Melke, J.; Schoekel, A.; Gerteisen, D.; Dixon, D.; Ettingshausen, F.; Cremers, C.; Roth, C.; Ramaker, D.E. Electrooxidation of ethanol on Pt. An in situ and time-resolved XANES study. *J. Phys. Chem. C* **2012**, *116*, 2838–2849. [CrossRef]
7. Jing, M.; Jiang, L.; Yi, B.; Sun, G. Comparative study of methanol adsorption and electro-oxidation on carbon-supported platinum in acidic and alkaline electrolytes. *J. Electroanal. Chem.* **2013**, *688*, 172–179. [CrossRef]
8. Da Silva, S.; Assumpção, M.; De Souza, R.; Buzzo, G.S.; Spinacé, E.V.; Neto, A.O.; Silva, J.C.M. Electrochemical and fuel cell evaluation of PtIr/C electrocatalysts for ethanol electrooxidation in alkaline medium. *Electrocatalysis* **2014**, *5*, 438–444. [CrossRef]
9. Silva-Junior, L.C.; Maia, G.; Passos, R.; De Souza, E.A.; Camara, G.; Giz, M.J. Analysis of the selectivity of PtRh/C and PtRhSn/C to the formation of CO2 during ethanol electrooxidation. *Electrochim. Acta* **2013**, *112*, 612–619. [CrossRef]
10. Comignani, V.; Sieben, J.M.; Brigante, M.; Duarte, M.M. Carbon supported Pt–NiO nanoparticles for ethanol electro-oxidation in acid media. *J. Power Sources* **2015**, *278*, 119–127. [CrossRef]
11. Antoniassi, R.M.; Silva, J.C.M.; Oliveira Neto, A.; Spinacé, E.V. Synthesis of Pt+ SnO_2/C electrocatalysts containing Pt nanoparticles with preferential (100) orientation for direct ethanol fuel cell. *Appl. Catal. B Environ.* **2017**, *218*, 91–100. [CrossRef]
12. Meenakshi, S.; Nishanth, K.G.; Sridhar, P.; Pitchumani, S. Spillover effect induced Pt-TiO2/C as ethanol tolerant oxygen reduction reaction catalyst for direct ethanol fuel cells. *Electrochim. Acta* **2014**, *135*, 52–59. [CrossRef]
13. Gudko, O.E.; Lastovina, T.A.; Smirnova, N.; Guterman, V.E.; Vladimir, G. Binary Pt-Me/C nanocatalysts: Structure and catalytic properties toward the oxygen reduction reaction. *Nanotechnol. Russ.* **2009**, *4*, 309–318. [CrossRef]
14. Brimaud, S.; Pronier, S.; Coutanceau, C.; Leger, J.-M. New findings on CO electrooxidation at platinum nanoparticle surfaces. *Electrochem. Commun.* **2008**, *10*, 1703–1707. [CrossRef]
15. Brunauer, S.; Emmett, P.H.; Teller, E. Adsorption of gases in multimolecular layers. *J. Am. Chem. Soc.* **1938**, *60*, 309–319. [CrossRef]
16. Watt-Smith, M.J.; Friedrich, J.; Rigby, S.P.; Ralph, T.R.; Walsh, F.C. Determination of the electrochemically active surface area of Pt/C PEM fuel cell electrodes using different adsorbates. *J. Phys. D Appl. Phys.* **2008**, *41*, 174004. [CrossRef]
17. Pang, H.; Lu, J.; Chen, J.; Chuang, Y.-C.; Liu, B.; Zhang, X. Preparation of SnO2-CNTs supported Pt catalysts and their electrocatalytic properties for ethanol oxidation. *Electrochim. Acta* **2009**, *54*, 2610–2615. [CrossRef]
18. Moghaddam, R.B.; Pickup, P.G. Support effects on the oxidation of ethanol at Pt nanoparticles. *Electrochim. Acta* **2012**, *65*, 210–215. [CrossRef]
19. Higuchi, E.; Miyata, K.; Takase, T.; Inoue, H. Ethanol oxidation reaction activity of highly dispersed Pt/SnO_2 double nanoparticles on carbon black. *J. Power Sour.* **2011**, *196*, 1730–1737. [CrossRef]
20. Gharibi, H.; Sadeghi, S.; Golmohammadi, F. Electrooxidation of Ethanol on highly active and stable carbon supported $PtSnO_2$ and its application in passive direct ethanol fuel cell: Effect of tin oxide synthesis method. *Electrochim. Acta* **2016**, *190*, 1100–1112. [CrossRef]

21. Tayal, J.; Rawat, B.; Basu, S. Bi-metallic and tri-metallic Pt–Sn/C, Pt–Ir/C, Pt–Ir–Sn/C catalysts for electro-oxidation of ethanol in direct ethanol fuel cell. *Int. J. Hydrog. Energy* **2011**, *36*, 14884–14897. [CrossRef]
22. Leontyev, I.; Leontyeva, D.V.; Kuriganova, A.; Popov, Y.V.; Maslova, O.A.; Glebova, N.; Nechitailov, A.A.; Zelenina, N.K.; Tomasov, A.; Hennet, L.; et al. Characterization of the electrocatalytic activity of carbon-supported platinum-based catalysts by thermal gravimetric analysis. *Mendeleev Commun.* **2015**, *25*, 468–469. [CrossRef]

Article

Sorbent Based on Polyvinyl Butyral and Potassium Polytitanate for Purifying Wastewater from Heavy Metal Ions

Anna Ermolenko [1], Maria Vikulova [1], Alexey Shevelev [1], Elena Mastalygina [2,3], Peter Ogbuna Offor [4], Yuri Konyukhov [5], Anton Razinov [2], Alexander Gorokhovsky [1] and Igor Burmistrov [1,2,5,*]

1 Department of Chemistry and Technology of Materials, Yuri Gagarin State Technical University of Saratov, 77 Polytecnicheskaya Street, 410054 Saratov, Russia; molish01@mail.ru (A.E.); vikulovama@yandex.ru (M.V.); titans5@rambler.ru (A.S.); algo54@mail.ru (A.G.)

2 Educational and Scientific Center "Trade", Plekhanov Russian University of Economics, 36 Stremyanny Lane, 117997 Moscow, Russia; elena.mastalygina@gmail.com (E.M.); razinov.ae@rea.ru (A.R.)

3 Emanuel Institute of Biochemical Physics, Russian Academy of Sciences, 4 Kosygina str., 119334 Moscow, Russia

4 Metallurgical and Materials Engineering Department, University of Nigeria, Nsukka 410001, Nigeria; peter.offor@unn.edu.ng

5 Department of Functional Nanosystems and High-Temperature Materials, National University of Science & Technology "MISIS", 4 Leninsky pr., 119049 Moscow, Russia; ykonukhov@misis.ru

* Correspondence: glas100@yandex.ru; Tel.: +7-91-7201-8703

Received: 14 May 2020; Accepted: 9 June 2020; Published: 13 June 2020

Abstract: Currently, the rapid development of industry leads to an increase in negative anthropogenic impacts on the environment, including water ecosystems. This circumstance entails toughening environmental standards and, in particular, requirements for the content of pollutants in wastewater. As a result, developing technical and cost-effective ways for wastewater purification becomes relevant. This study is devoted to the development of a novel composite sorbent, based on polyvinyl butyral and potassium polytitanate, designed to purify water from heavy metal ions. The co-deposition of a mixture based on a polymer solution and a filler suspension was used to obtain a composite material. In this work, the influence of the deposition conditions on the structure and properties of the resulting composites was studied, as well as the optimal ratio of components, including solvent, precipitant, polymer binder, and filler, were established. In the course of this study on the sorption properties of the developed composite materials using various sorption models, the sorption capacity of the obtained material, the sorption mechanism, and the limiting stage of the sorption process were determined. The developed sorbent can be suitably used in the wastewater treatment systems of galvanic industries, enterprises producing chemical current sources, and in other areas.

Keywords: polymer composite; sorbent; wastewater; polyvinyl butyral; potassium polytitanate; heavy metals

---

1. Introduction

The increase in environmental pollution is one of the main problems of modern society. The number of various pollutants, including heavy metal ions [1], toxicants and dyes [2,3], bacteria, and viruses, entering the wastewater system is rapidly increasing due to the rapid development of industry and population growth. In this regard, the problem of water purification becomes more relevant, which is reflected in the number of studies devoted to this topic.

Heavy metal ions are not destroyed under environmental factors, and, on the contrary, they tend to accumulate in various living organisms. Getting involved in complex food webs, heavy metals adversely affect plants and animals and end up in the human diet. Owing to this, the pollution of wastewater by heavy metal ions is a problem of increasing importance for environmental, evolutionary, food, and environmental reasons [4].

There is a wide variety of ways employed in treating wastewater for heavy metal ions, including chemical precipitation [5], ion exchange [6], reverse osmosis [7], electrolysis [8], membrane filtration [9], and adsorption [10–13]. The listed methods have both advantages and disadvantages. Among the technologies implemented in practice, ion-exchange sorption is one of the most effective and cost-effective methods of removing heavy metals from aqueous media [14–16]. This water treatment method provides a high degree of purification (close to 100%) for highly polluted water. The possibility of sorbent regeneration can be realized with a relatively simple hardware design. The most common sorption materials for heavy metals are synthetic zeolite, natural kaolinite, chitin, chitosan [17–21], carbon materials [22–24], and agricultural waste, such as rice bran and orange peels [25]. According to recent studies, synthetic nanostructured carbon sorbents [18,26] have a very high sorption capacity (up to 97 mg of Pb^{2+} per gram of sorbent) [27], which makes them one of the most interesting objects of research in this field. Nevertheless, there are several disadvantages of carbon sorbents that limit their widespread and routine use. The disadvantages include high cost, as well as limited volumes and the complexity of waste product disposal [28]. Natural sorbents based on layered aluminosilicates (kaolinite, montmorillonite, zeolite, and their analogues) are quite cheap and widespread. However, their sorption properties depend significantly on many external factors, such as deposits and impurities, pH level, temperature, the presence of either or both electrolytes and surfactants, etc. [13,29]. Hence, the need for the complex chemical and thermal modification of the feedstock before practical use.

Among synthetic sorption materials with a layered structure, potassium titanates show a high sorption capacity, and good sintering into glass–ceramic composites, whereby they can be effectively processed into safe and valuable ceramic products [30]. The sorption characteristics of some crystalline modifications of potassium titanate were studied in the works [31,32]. For instance, the sorption capacity of crystalline potassium hexatitanate ($K_2Ti_6O_{13}$) for Cd^{2+} ions can be up to 0.80 mmol/g [31], and the sorption capacity of crystalline potassium tetratitanate ($K_2Ti_4O_9$) for Cu^{2+} ions can be up to 1.94 mmol/g [32]. Amorphous potassium titanates demonstrated effectiveness in the sorption of organic dyes [33] and Ni^{2+} ions [34]. In our previous work, the sorption of Pb^{2+} and Sr^{2+} ions by the amorphous potassium titanate was studied. It was shown that the sorption capacity for Pb^{2+} ions was 714 mg/g and for Sr^{2+} ions it was 345 mg/g [35].

To avoid the problem of separating the filtrate from the purified solution for powder catalysts, a number of techniques can be used, for example, granulation, coating on substrates, creating membranes, and the formation of polymer composites. In order to develop composite sorbents, carbon nanotubes, lignin, graphene oxide, chitosan, concrete, and others, are used. When creating composites with sorption properties, a number of difficulties arise, including the dispersion of filler in a polymer matrix and the filler pre-treatment, which ensures the availability of particles of the filler (sorbent) for ions from the solution.

Previously, polyvinyl butyral (PVB) was studied as a polymer matrix for obtaining composites with sorption properties [30]. Nevertheless, the sorption composite was used in the form of a thin-film coating, which is convenient for photocatalytic systems studied by the authors but is ineffective for the absorption of soluble ions from the water stream.

In this regard, the aims of this study were to create a composite based on potassium polytitanate and polyvinyl butyral with open porosity for the effective sorption of heavy metal ions from a solution and to analyze its sorption capacity and mechanism. These aims were achieved by creating a porous highly filled composite by a coagulation precipitation technique previously used for polypropylene-based composites [36]. Isopropyl alcohol was used as a solvent for PVB, and distilled water was used as a precipitant. The dependence of the structure of the composites and properties on the deposition conditions were studied. The sorption mechanism was determined for the composites with an optimal structure (large open porosity, high degree of filling).

2. Materials and Methods

2.1. Materials

Potassium polytitanates (PPT) were synthesized in a hydroxide-salt melt, according to the procedure described in [37]. As the raw components, titanium dioxide TiO_2 (99% anatase, Aldrich, Darmstadt, Germany, CAS 13463-67-7), potassium hydroxide KOH (98% purity, Vekton, St. Petersburg, Russia), and potassium nitrate KNO_3 (98% purity, Reahim, Moscow, Russia), in a mass ratio of 30:30:40, were used. The components were blended with distilled water in a ratio of water/titanium dioxide = 2:1, then exposed to thermal treatment in a muffle furnace for 3 h at a temperature of 500 °C.

Polyvinyl butyral (hereinafter PVB), of the film-cast type (PP grade, GOST 9439-85) and supplied by "VitaKhim" LLC (Moscow, Russia), was used as a polymer binder for the composites. To prepare the PVB solution, the dehydrated isopropyl alcohol (hereafter IPA) (GOST 9805-84), supplied by "Zavod sinteticheskogo spirta" CJSC (Orsk, Russia), was used.

Standardized test solutions for the study of sorption capacity were made of lead nitrate (high grade, GOST 4236-77, Russia) and manufactured by NPF "Nevsky Khimik" (St. Petersburg, Russia).

2.2. Preparing a Highly Filled Porous Composite Based on Potassium Polytitanates

PVB was dissolved in isopropyl alcohol at an elevated temperature of 90 °C by a mechanical stirrer operating at a speed of 150 rpm. Simultaneously, a suspension of PPT in isopropyl alcohol was prepared by stirring with a mechanical stirrer (200 rpm) in a ratio of 1, 3, or 5 g of PPT to 50 mL of alcohol (in order to ensure the different filling degree of composites). After dissolving PVB, the solution was combined with the suspension of PPT (to obtain the concentrations shown in Table 1) and mixed for 15 min by a mechanical stirrer (150 rpm). Then the resulting mass was poured (thin stream) into an excess of precipitant (water) where rapid coagulation occurred. The volume ratio of the PVB solution (mixed with PPT powder) to the precipitant (water) was 1:2, 1:4, or 1:6. After formation, the composite was removed from the water–alcohol solution by decantation and dried at 20 °C for 24 h.

Table 1. The compositions of porous composite materials based on potassium polytitanate.

N°	Sample Notation	PVB * Content in Composite, wt.%	PPT ** Content in Composite, wt.%	Ratio of Isopropyl Alcohol/Water
1	1/6-70-30	70	30	1/6
2	1/4-70-30	70	30	1/4
3	1/2-70-30	70	30	1/2
4	1/4-50-50	50	50	1/4
5	1/4-90-10	90	10	1/4

* Polyvinyl butyral. ** Potassium polytitanates.

By a process of rapid coagulation of PVD, when alcohol is mixed with water, a sponge-like composite was formed. The type of porosity of the resulting composite depended on the volume ratio of the suspension to the precipitant, as well as the content of PPT in the initial suspension. A wide range of PPT concentrations and several different ratios of isopropyl alcohol/water were studied to provide the maximum concentration of PPT (because only PPT sorbs heavy metal ions, but PVB is just a passive matrix holding titanate) and to keep some strength of the sponge-like composite (so that the PPT-based sorbent was not crumbled during operation, cutting, or pouring into different containers).

The surface morphology of the composites was studied by a US-made Aspex Explorer scanning electron microscope (SEM) with an energy dispersion attachment.

2.3. Sorption Capacity Assessment and Kinetic Analysis

To determine the dynamic sorption characteristics of the composites, the installation shown in Figure 1 was used.

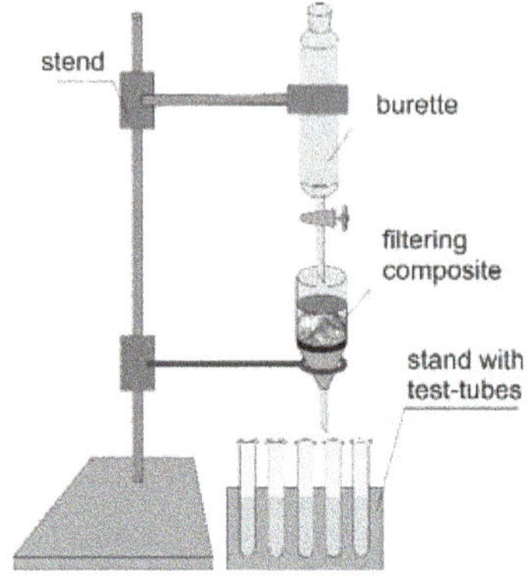

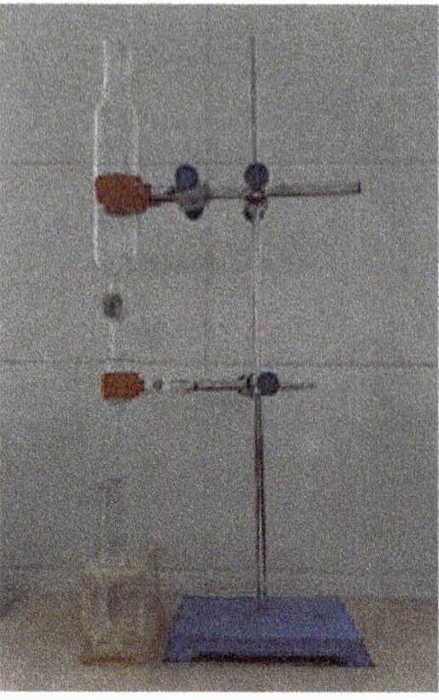

Figure 1. Installation for the study of the process of the dynamic sorption of lead ions.

The solution of lead nitrate (75 g/L), with a constant speed (6.5 mL/min), was fed into the funnel, in which there was the porous polymer composite. The filtrate was collected in 10 mL portions. Furthermore, the obtained materials were studied using X-ray fluorescence analysis, which was carried out using a Spectroscan-MAX-G spectrometer (LLC "NPO" SPECTRON, Russia, St. Petersburg) with a scanning crystal-diffraction channel. Gauge curves were previously found to determine the concentration.

The values obtained by changes in the concentration of lead ions over time were evaluated using the following models: Boyd model, Lagergren pseudo-first order kinetic model, Ho and McKay's pseudo-second order kinetic model, and Weber–Morris intraparticle diffusion model.

The Boyd model allows the evaluation of the role of diffusion in the sorption process. For this model, development Equation (1) was used [38]:

$$F = \frac{Q_t}{Q_e} = 1 - \left(\frac{6}{\pi^2}\right)\exp(-Bt), \quad (1)$$

where F = degree of process completion, Q_t = quantities of ions at current time (mg/g), Q_e = quantities of ions in a state of sorption quasi-equilibrium (mg/g), which are adsorbed at time t and have a state of sorption quasi-equilibrium, respectively.

The Lagergren Equation (2) describes the pseudo-first order kinetic model for sorption processes [39]:

$$\text{Lg}(Q_e - Q_t) = \text{lg}Q_e - \frac{k_1}{2303}t, \quad (2)$$

where k_1 = rate constant (min^{-1}).

To consider the kinetics of the sorption process by a pseudo-second order model, a Ho and McKay Equation (3) was used [40]:

$$\frac{t}{Q_t} = \frac{1}{k_2 Q_e^2} + \frac{t}{Q_e}, \quad (3)$$

where k_2 = rate constant (g·mg^{-1}·min^{-1}).

A Weber–Morris intraparticle diffusion model was expressed by the following Equation (4) [41]:

$$q_t = K_{dif} t^{0.5} + C_i, \tag{4}$$

where K_{dif} = intraparticle diffusion rate constant (mg·g^{-1}·min$^{-0.5}$), and C_i = parameter characterizing the thickness of the boundary layer (mg/g).

3. Results and Discussion

To determine the effects of the volume ratio of the suspension/precipitant and PPT content in the initial suspension on the composite structure, a range of samples were used (Table 1). The choice was based on the desire to create a composite with a maximum concentration of PPT, but at the same time, so that it did not crumble during operation, cutting, or pouring into different containers.

The appearance of the cross-section of the studied composite is shown in Figure 2. The cross-section was obtained manually by slicing blades. There was a significant decrease in the porosity of the composite materials (Figure 2a–c), that could be caused by a decrease in the amount of precipitant. When the isopropyl alcohol/water ratio was 1:2 (Figure 2c), most of the internal pores were insulated (almost no open porosity remained) and, in turn, the outer surface of the composite became smooth. The composite sample 1/4-50-50 (Figure 2d) had a low porosity compared to the sample 1/2-70-30. In this case, this is due to the high filler content (mass ratio filler/matrix = 1:1). In the last studied composite (Figure 2e), the amount of potassium polytitanate introduced into the PVB was the smallest, which led to the immersion and isolation of PPT particles in the composite matrix, and as a result of which their activity, the surface was not available for physicochemical processes.

(a) (b) (c) (d) (e)

Figure 2. The slices of the obtained porous composite materials: (**a**) 1/6-70-30; (**b**) 1/4-70-30; (**c**) 1/2-70-30; (**d**) 1/4-50-50; (**e**) 1/4-90-10.

Figure 3 presents microphotographs of the composite samples obtained by scanning electron microscopy (SEM). The filling degree and the conditions for obtaining them significantly affected the porosity of the materials. The composite containing 10 wt.% of PPT (sample 1/4-90-10) had a dense structure with a predominantly closed porosity. When 50 wt.% of the filler was introduced into the polymer matrix, the composite was characterized by a loose structure that could be easily destroyed by the slightest mechanical stress.

With a change in the volume ratio of the PPT suspension to the precipitant, the pattern of micropores changed with a change in macropores (Figure 3). For the first two samples (1/6-70-30 and 1/4-70-30), the pore size was 20–30 μm and 5–10 μm, respectively. The third sample (1/2-70-30) was characterized by a significant decrease in the number of micropores and the formation of closed pores (size did not exceed 5 μm). For sample 1/6-70-30, the presence of PPT particles localized in pores and a lack of moistening with a binder was observed (Figure 3a). Moreover, an increase in the PTT content in the initial suspension (sample 1/4-50-50) led to the predominant localization of PPT in the form of individual particles that did not interact with the binder. The observed distribution of PPT particles could provide the most effective sorption, but, obviously, could lead to the chipping of the sorption filler from the composite.

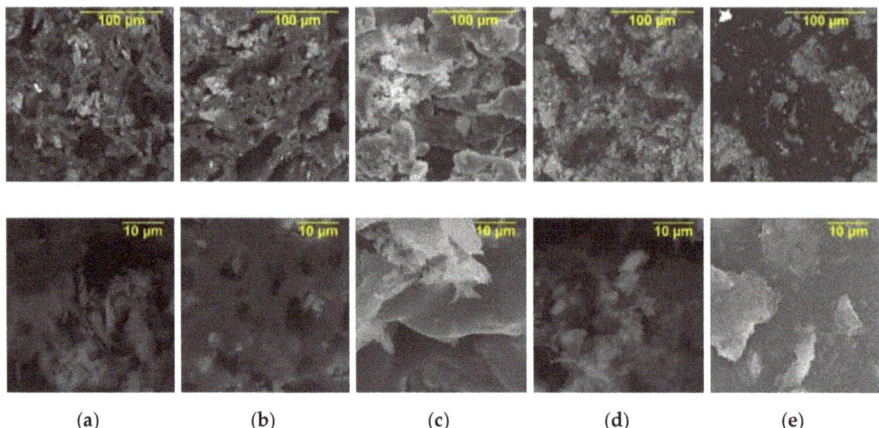

Figure 3. The microphotographs of the composites obtained by SEM: (**a**) sample 1/6-70-30 based on 70 wt.% of PVB and 30 wt.% of PPT, IPA/water = 1:5; (**b**) sample 1/4-70-30 based on 70 wt.% of PVB and 30 wt.% of PPT, IPA/water = 1:4; (**c**) sample 1/2-70-30 based on 70 wt.% of PVB and 30 wt.% PPT, IPA/water = 1:2; (**d**) sample 1/4-50-50 based on 50 wt.% of PVB and 50 wt.% of PPT, IPA/water = 1:4; (**e**) sample 1/4-90-10 based on 90 wt.% of PVB and 10 wt.% of PPT, IPA/water = 1/4. Abbreviations: SEM—scanning electron microscopey; PVB—polyvinyl butyral; PPT—potassium polytitanates; IPA—isopropyl alcohol.

The reduction in the amount PPT in the initial suspension (sample 1/4-90-10) led to a noticeable aggregation (most likely due to the increased viscosity of the suspension). While the filler particles were completely wetted by the matrix, which could increase the mechanical properties of the materials, this could limit access to filtered solutions and adversely affect sorption characteristics.

The sample 1/4-90-10 obviously cannot be an effective sorption material, primarily because of the low content of the active component in the composite. In addition, PPT particles were almost completely immersed in the polymer matrix, which blocked them from interacting with heavy metal ions. The composite 1/4-50-50, on the contrary, contained the largest amount of filler. However, its further use was limited by a low degree of fixation of PPT particles by the polymer binder, which created the probability of the dispersed powder getting into the purified solution, and, in turn, the necessity of the additional cleaning of the solution. The samples with a ratio of PPT/PVB = 30:70 were considered optimal, since the concentration of an active component was sufficient for effective sorption interaction with an optimal ratio with a binder, which ensured the effective fixation of particles of potassium polytitanate, while maintaining the maximum available surface area. However, the high viscosity of the solvent in the case of a dilution of 1/4 and 1/2 had a significant effect on the micropore size, reducing them by more than three to four times, which was one of the determining factors for evaluating the sorption ability.

According to the study of the composite structure, it was assumed that the sample 1/6-70-30, containing 70 wt.% of PVB and 30 wt.% of PPT (IPA/water = 1:6), had the best sorption ability. Therefore, the sorption of lead ions by this material was studied under dynamic conditions.

It is necessary to give an isotherm describing the dependence of the concentration on time and show the sorption capacity. The effect of the contact time in the range of 1.5–32.0 min on the lead ion sorption by composite materials is shown in Figure 4.

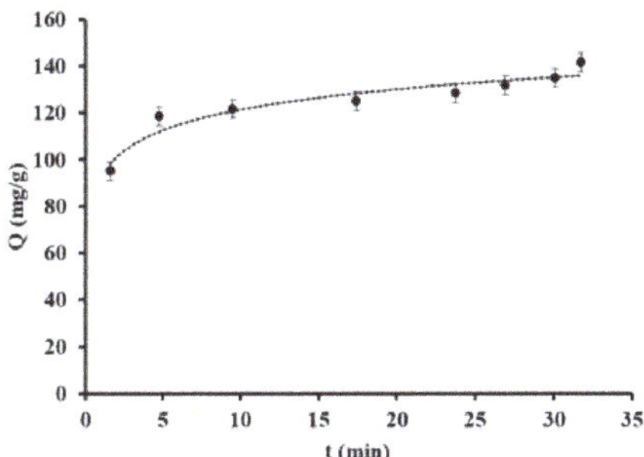

Figure 4. Effect of contact time on the lead ion sorption by the porous polymer composite of 1/6-70-30.

According to the kinetics curve, the sorption of lead ions by the studied polymer composite was a quick process. After 5 min of interaction with the polymer composite, most of the lead ions were removed from the solution. A highly concentrated solution (0.23 mol/L) and a high filtration rate (6.5 mL/min) were used to study the sorption kinetics. Under these conditions, the residual concentration was 0.09 mol/L. The sorption capacity of the composite began to increase slowly when the time increased from 10 to 30 min. The porous structure of the composite and available active sites of potassium polytitanate provided the high sorption rate at the beginning of the process. When the surface of the sorbent was covered and blocked, the sorption capacity value stopped changing. A little more than 30 min was needed to attain the equilibrium with a sorption capacity of 141.5 mg/g.

The role of diffusion during sorption was evaluated using the Boyd model. Figure 5a shows the time-varying behavior of the parameter –ln(1-F), which estimates the contribution of external diffusion to the overall process rate during the sorption of lead ions.

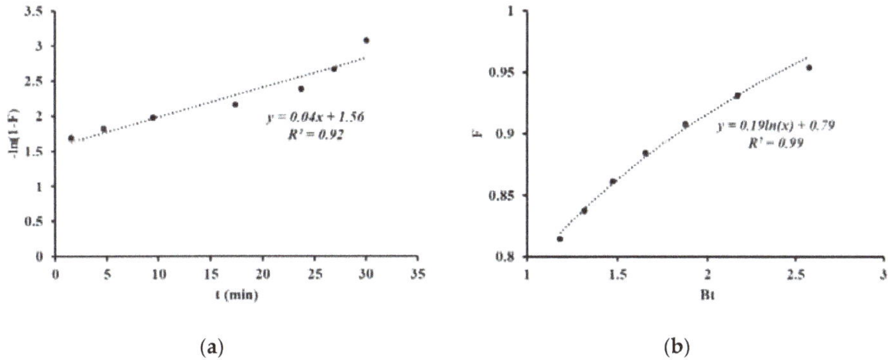

Figure 5. The dependence of –ln(1-F) on time (t) (**a**) and sorption degree (F) on Bt parameter (**b**) for the sorption of lead ions by the porous polymer composite of 1/6-70-30.

According to the calculated coefficient of determination (R^2 ~ 0.9), during the sorption of lead ions by the porous composite, the external processes occurred, but they were not limiting. It should be noted that the plot of the dependence of the –ln(1-F) parameter on time did not pass through the origin of coordinates, which indicates the mechanism of the mixed diffusion of sorption.

In Figure 5b, the contribution of an internal diffusion to the sorption rate was estimated by the dependence of F–Bt. The high coefficient of determination (R^2) of the existing dependence between the experimental and calculated data of the logarithmic trend line indicates the high role of internal diffusion during the sorption of lead ions by the porous polymer composite.

The Weber–Morris model takes into account the diffusion at the phase boundary and inside the sorbent particles. Figure 6 shows the kinetics of sorption defined according to this model, using Equation (4).

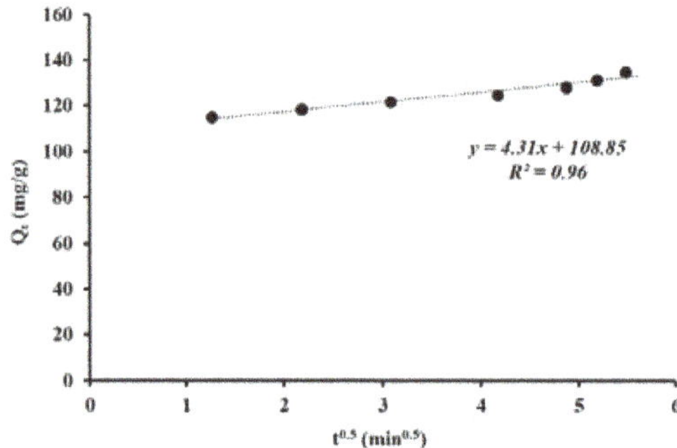

Figure 6. The kinetics of sorption by the porous polymer composite of 1/6-70-30 in the coordinates of the Weber–Morris model.

The existing dependence of Q_t on $t^{0.5}$ fits well on the straight line, which is described by the equation y = 4.31x + 108.85. It indicates that the rate of the sorption process was determined by the diffusion inside the sorbent particles. Thus, composite 1/6-70-30 had enough macroporosity and microporosity to provide an effective sorption, limited by the processes inside the aggregates of PPT particles, not by the rate of access to them. However, the linear plots did not pass through the origin. Such behavior of line could be due to the difference in the rate of mass transfer in the various sorption stages and some degree of boundary layer limits. Based on the obtained equation of linear dependence, diffusion parameters were calculated. The parameter of the boundary layer thickness (C_i) was equal to 108.85 mg/g, which confirmed the rapid sorption in a short period. The value of the intraparticle diffusion rate constant (K_{dif} = 4.31 mg·g^{-1}·min$^{-0.5}$) indicates a rapid diffusion process.

The chemical kinetics models of pseudo-first and pseudo-second orders describe the contribution of chemical reactions to the sorption rate. The correspondence of the sorption processes to the pseudo-first order kinetic model was estimated using the Lagergren Equation (2). Figure 7 shows the kinetics of sorption, according to the pseudo-first (Figure 7a) and the pseudo-second (Figure 7b) order models.

Since the coefficient of determination was lower than 0.95, the contribution of diffusion at a stage not preceding sorption was not significant. The pseudo-second order model was constructed using the Ho and McKay Equation (3). The high value of the determination coefficient indicates the correspondence of this process to the pseudo-second order model and can be described by the equation y = 0.0074x + 0.0057, by which the kinetic parameters were determined. The implication is that the stage of chemical interaction of lead ions and PPT is a limiting one during the sorption (k_2 = 0.006 g·mg^{-1}·min^{-1}, Q_e = 137.0 mg/g). According to this model, the interaction of the sorbent and sorbate obeys the law of the mass action. Therefore, the rate of a chemical reaction is directly proportional to the concentration of two substances reacting in a ratio of 1:1.

Table 2 shows the comparison of the sorption capacity of the developed polymer composite based on polyvinyl butyral and potassium polytitanate (sample 1/6-70-30) with the same parameter as other polymer composites with respect to lead ions. According to the data, the sorption capacity of the developed sorbent based on polyvinyl butyral and potassium polytitanate is third only to those of composites based on poly (3,4-ethylenedioxythiophene)/lignin and poly (N-vinylcarbazole)/graphene oxide.

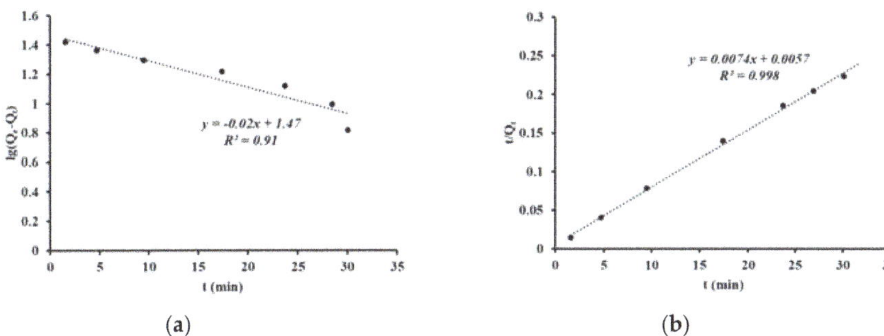

Figure 7. The kinetics of sorption by the porous polymer composite of 1/6-70-30 in the coordinates of the pseudo-first (**a**) and (**b**) the pseudo-second order models.

In the case of the composite obtained by Checkol F. et al. [42], at a high efficiency, even after 60 cycles (47% of the sorption capacity was retained), obtaining a composite film was rather labor-intensive by galvanostatic deposition. The degradation of the composite developed by Karnib M. et al. [43] was not studied, while the necessity of activating the filler surface by heat treatment and drying in a vacuum was indicated, which requires additional equipment. Vacuum drying was also required when manufacturing the composites in the other research [44,45]. The rest of the composites are characterized by a low sorption capacity to lead ions in comparison with the material developed in this work. Moreover, these composites are characterized by complex manufacturing technologies with preliminary functionalization of the filler surface [44–47], in situ polymerization [44], or using irradiation and a nitrogen atmosphere [48]. For a number of composites [44,46,47,49], the efficiency of removing lead ions after three to four cycles of the process did not decrease significantly, and for the others this parameter was not studied. In addition, the high cost of the composite materials due to the cost of polymers or fillers should be noted; in some cases, both components cost several thousand or tens of thousands of roubles.

Table 2. The sorption capacity of different polymer matrix composites with respect to lead ions.

Polymer Composite	Sorption Capacity (mg/g)	Reference
Polyvinyl butyral/potassium polytitanate (1/6-70-30 sample)	137.0	-
Potassium polytitanate	714.3	[35]
Poly (3,4-ethylenedioxythiophene)/lignin	452.8	[42]
Polypyrrole/multi-wall carbon nanotube	26.3	[44]
Poly (ethylene imine)/silica gels	82.6	[49]
Ion-imprinted polymer Pb(II)/SBA-15	42.6	[46]
Poly-methacrylic acid grafted chitosan-bentonite nanocomposite	111.0	[48]
Cellulose/TiO_2	120.5	[47]
Carboxymethyl-β-cyclodextrin/Fe_3O_4	64.5	[45]
Poly (N-vinylcarbazole)/graphene oxide	982.9	[43]

The most interesting activity is the comparison of the results obtained for the composite polyvinyl butyral/potassium polytitanate with the results for pure potassium polytitanate, which was studied in our previous work [33]. The sample 1/6-70-30 had the PVB/PPT ratio of 70/30. For this sample, if the PPT sorption capacity is fully realized (assuming that the PVB sorption capacity is negligible), the sorption capacity of the composite will be 30% of the PPT capacity (214 mg/g). However, according to the experimental data, the sample 1/6-70-30 was characterized by the sorption capacity being equal to 137.0 mg/g. Obviously, the polymer matrix hindered access to the PPT sorbent (filler), thus reducing the sorption capacity. Nevertheless, the sorption capacity of the developed polymer composite was about 70% of the sorption capacity of the pure PPT.

4. Conclusions

The polyvinyl butyral/potassium polytitanate composite sorbent for lead ions was investigated. The influence of the composite production conditions on the sorption properties of the developed composite materials was studied.

The sorption mechanism was determined using various kinetic models. The sorption process of lead ions by the polyvinyl butyral/potassium polytitanate composite was confirmed by a pseudo-second order kinetic model with a better correlation compared with the pseudo-first. It meant that both the concentrations of sorbate and sorbent were the factors determining the process rate. Additionally, this kinetic model showed that the potential rate-controlling stage of the sorption was a chemisorption with cation exchange and the participation of potassium cations in either or both the PPT interlayer space reaction and the complexation reaction on the sorbent surface. The calculated value of $Q_{e\,(calculated)}$ = 137.0 mg/g deviated slightly from the experimental one, $Q_{e\,(experimental)}$ = 141.5 mg/g. The conformity of the experimental data to diffusion models was evidence that the external mass transfer was of little importance. Hence, the intraparticle diffusion was significant in the sorption rate determination for the system of PVB/PPT–Pb(II), especially at the later stages of this process. A Boyd kinetic plot confirmed that the internal mass transfer was the slowest step involved in the sorption process.

Due to the influence of the polymer matrix, which impedes the access of lead ions from the solution to the sorbent (filler), the sorption capacity of polytitanate was just partially realized. When chemical particles interact freely and the optimal parameters needed for the preparation of the composites are in place, the sorption capacity of polytitanate can be up to 70% of the value of pure filler. Based on the results obtained, the polyvinyl butyral/potassium polytitanate composite sorbent can be used for the treatment systems of wastewater containing lead ions. For example, for treatment of wastewater from ore-dressing factories and mines, metallurgical plants, and many chemical industries, such as the for the production of batteries, glass, paints, insecticides, and gasoline, etc.

Author Contributions: Conceptualization, A.G. and I.B.; methodology, A.E.; software, Y.K. and A.R.; validation, I.B., A.S. and A.G.; formal analysis, A.G.; investigation, A.E., M.V. and A.S.; resources, A.G. and I.B.; data curation, A.S.; writing—original draft preparation, E.M. and P.O.O.; writing—review and editing, E.M. and P.O.O.; visualization, M.V. and A.R.; supervision, I.B.; project administration, Y.K. and A.S. All authors have read and agreed to the published version of the manuscript.

Funding: This research received no external funding.

Conflicts of Interest: The authors declare no conflict of interest.

References

1. Masindi, V.; Muedi, K.L. Environmental Contamination by Heavy Metals. In *Heavy Metals*; Hosam El-Din, M., Saleh, R.A., Eds.; IntechOpen: London, UK, 2018; pp. 1151–1235.
2. Wen, Y.; Schoups, G.; van de Giesen, N. Organic pollution of rivers: Combined threats of urbanization, livestock farming and global climate change. *Sci. Rep.* **2017**, *7*, 43289. [CrossRef] [PubMed]
3. Karpi, J.; Kotowska, U. Removal of Organic Pollution in the Water Environment. *Water* **2019**, *11*, 2017.
4. Nagajyoti, P.C.; Lee, K.D.; Sreekanth, T.V.M. Heavy metals, occurrence and toxicity for plants: A review. *Environ. Chem. Lett.* **2010**, *8*, 199–216. [CrossRef]
5. Wang, L.K.; Hung, Y.T.; Shammas, N.K. Chemical precipitation physicochemical treatment processes. *Hum. Press* **2005**, *3*, 141–197.
6. Xing, Y.; Chen, X.; Wang, D. Electrically regenerated ion exchange for removal and recovery of Cr (VI) from wastewater. *Environ. Sci. Technol.* **2007**, *41*, 1439–1443. [CrossRef]
7. Bakalar, T.; Bugel, M.; Gadoysova, L. Heavy metal removal using reverse osmosis. *Acta Montan. Slovaca* **2009**, *14*, 250–253.
8. Rahul, K.J. Application of electro-dialysis (ED) to remove divalent metals ions from wastewater. *Int. J. Chem. Sci. Appl.* **2013**, *4*, 68–72.

9. Ersahin, M.E.; Ozgun, H.; Dereli, R.K.; Ozturk, I.; Roest, K. A review on dynamic membrane filtration: Materials, applications and future perspectives. *Bioresource Technol.* **2012**, *122*, 196–206. [CrossRef]
10. Wanees, S.A.; Ahmed, A.M.M.; Adam, M.S.; Mohamed, M.A. Adsorption studies on the removal of hexavalent chromium-contaminated wastewater using activated carbon and bentonite. *Chem. J.* **2012**, *2*, 95–105. [CrossRef]
11. Vindoh, R.; Padmavathi, R.; Sangeetha, D. Separation of heavy metals from water samples using anion exchange polymers by adsorption process. *Desalination* **2011**, *267*, 267–276. [CrossRef]
12. Al-Degs, Y.S.; El-Barghouthi, M.I.; Issa, A.A.; Khraisheh, M.A.; Walker, G.M. Sorption of Zn(II), Pb(II) and Co(II) using natural sorbents: Equilibrium and kinetic studies. *Water Res.* **2006**, *40*, 2645–2658. [CrossRef] [PubMed]
13. Wu, P.; Wu, W.; Li, S. Removal of Cd^{2+} from aqueous solution by adsorption using Fe-montmorillonite. *J. Hazard. Mater.* **2009**, *169*, 824–830. [CrossRef] [PubMed]
14. Ali, I.; Gupta, V.K. Advances in water treatment by adsorption technology. *Nat. Protoc.* **2006**, *1*, 2661–2667. [CrossRef] [PubMed]
15. Jiuhui, Q.U. Research progress of novel adsorption processes in water purification. *J. Environ. Sci.* **2008**, *20*, 1–13.
16. Shahmirzadi, M.A.A.; Hosseini, S.S.; Luo, J.; Ortiz, I. Significance, evolution and recent advances in adsorption technology, materials and processes for desalination, water softening and salt removal. *J. Environ. Manag.* **2018**, *215*, 324–344. [CrossRef]
17. Bhattacharyya, K.G.; Gupta, S.S. Adsorption of a few heavy metals on natural and modified kaolinite and montmorillonite: A review. *Adv. Colloid Interface Sci.* **2008**, *140*, 114–131. [CrossRef]
18. Płaza, A.; Kołodyńska, D.; Hałas, P.; Gęca, M.; Franus, M.; Hubicki, Z. The zeolite modified by chitosan as an adsorbent for environmental applications. *Adsorp. Sci. Technol.* **2017**, *35*, 834–844. [CrossRef]
19. Karthikeyan, G.; Andal, N.M.; Anbalagan, K. Adsorption studies of iron (III) on chitin. *J. Chem. Sci.* **2005**, *117*, 663–672. [CrossRef]
20. Karthikeyan, G.; Anbalagan, K.; Andal, N.M. Adsorption dynamics and equilibrium studies of Zn (II) onto chitosan. *J. Chem. Sci.* **2004**, *116*, 119–127. [CrossRef]
21. Wu, F.-C.; Tseng, R.-L.; Juang, R.-S. A review and experimental verification of using chitosan and its derivatives as adsorbents for selected heavy metals. *J. Environ. Manag.* **2010**, *91*, 798–806. [CrossRef]
22. Rao, G.P.; Lu, C.; Su, F. Sorption of divalent metal ions from aqueous solution by carbon nanotubes: A review. *Sep. Purif. Technol.* **2007**, *58*, 224–231. [CrossRef]
23. Xue, C.; Qi, P.; Liu, Y. Adsorption of aquatic Cd^{2+} using a combination of bacteria and modified carbon fiber. *Adsorp. Sci. Technol.* **2018**, *36*, 857–871. [CrossRef]
24. Guo, T.; Bulin, C.; Li, B.; Zhao, Z.; Yu, H.; Sun, H.; Ge, X.; Xing, R.; Zhang, B.; Zhang, B. Efficient removal of aqueous Pb(II) using partially reduced graphene oxide-Fe_3O_4. *Adsorp. Sci. Technol.* **2018**, *36*, 1031–1048. [CrossRef]
25. O'Connell, D.W.; Birkinshaw, C.; O'Dwyer, T.F. Heavy metal adsorbents prepared from the modification of cellulose: A review. *Bioresour. Technol.* **2008**, *99*, 6709–6724. [CrossRef]
26. Wang, X.; Guo, Y.; Yang, L.; Han, M.; Zhao, J.; Cheng, X. Nanomaterials as sorbents to remove heavy metal ions in wastewater treatment. *J. Environ. Anal. Toxicol.* **2012**, *2*, 154–158. [CrossRef]
27. Li, Y.H.; Ding, J.; Luan, Z.; Di, Z.; Zhu, Y.; Xu, C.; Wu, D.; Wei, B. Competitive adsorption of Pb^{2+}, Cu^{2+} and Cd^{2+} ions from aqueous solutions by multiwalled carbon nanotubes. *Carbon* **2003**, *41*, 2787–2792. [CrossRef]
28. Smart, S.K.; Cassady, A.I.; Lu, G.Q.; Martin, D.J. The biocompatibility of carbon nanotubes. *Carbon* **2006**, *44*, 1034–1047. [CrossRef]
29. Gupta, S.S.; Bhattacharyya, K.G. Adsorption of heavy metals on kaolinite and montmorillonite: A review. *Phys. Chem. Chem. Phys.* **2012**, *14*, 6698–6723. [CrossRef]
30. Gorokhovsky, A.; Vikulova, M.; Escalante-Garcia, J.I.; Tretyachenko, E.; Burmistrov, I.; Kuznetsov, D.; Yuri, D. Utilization of nickel-electroplating wastewaters in manufacturing of photocatalysts for water purification. *Process Saf. Environ. Prot.* **2020**, *134*, 208–216. [CrossRef]
31. Mishra, S.P.; Singh, V.K.; Tiwari, D. Radiotracer technique in adsorption study: Part XVII. Removal Behaviour of Alkali Metal (K- and Li-) Titanates for Cd (II). *Appl. Radiat. Isot.* **1998**, *49*, 1467–1475. [CrossRef]

32. Nunes, L.M.; Cardoso, V.A.; Airoldi, C. Layered titanates in alkaline, acidic and intercalated with 1,8-octyldiamine forms as ion-exchangers with divalent cobalt, nickel and copper cations. *Mater. Res. Bull.* **2006**, *41*, 1089–1096. [CrossRef]
33. Tretyachenko, E.V.; Gorokhovsky, A.V.; Yurkov, G.Y.; Fedorov, F.S.; Vikulova, M.A.; Kovaleva, D.S.; Orozaliev, E.E. Adsorption and photo-catalytic properties of layered lepidocrocite-like quasi-amorphous compounds based on modified potassium polytitanates. *Particuology* **2014**, *17*, 22–28. [CrossRef]
34. Gorokhovsky, A.V.; Tret'yachenko, E.V.; Vikulova, M.A.; Kovaleva, D.S.; Yurkov, G.Y. Effect of Chemical Composition on the Photocatalytic Activity of Potassium Polytitanates Intercalated with Nickel Ions. *Russ. J. Appl. Chem.* **2013**, *86*, 343–350. [CrossRef]
35. Ermolenko, A.; Shevelev, A.; Vikulova, M.; Blagova, T.; Altukhov, S.; Gorokhovsky, A.; Godymchuk, A.; Burmistrov, I.; Offor, P.O. Wastewater Treatment from Lead and Strontium by Potassium Polytitanates: Kinetic Analysis and Adsorption Mechanism. *Processes* **2020**, *8*, 217. [CrossRef]
36. Mazov, I.; Burmistrov, I.; Il'Inykh, I.; Stepashkin, A.; Kuznetsov, D.; Issi, J.-P. Anisotropic thermal conductivity of polypropylene composites filled with carbon fibers and multiwall carbon nanotubes. *Polym. Compos.* **2015**, *36*, 1951–1957. [CrossRef]
37. Sanchez-Monjaras, T.; Gorokhovsky, A.; Escalante-Garcia, J.I. Molten salt synthesis and characterization of potassium polytitanate ceramic precursors with varied TiO_2/K_2O molar ratios. *J. Am. Ceram. Soc.* **2008**, *91*, 3058–3065. [CrossRef]
38. Boyd, G.E.; Adamson, A.W.; Myers Jr, L.S. The exchange adsorption of ions from aqueous solutions by organic zeolites. II. Kinetics1. *J. Am. Chem. Soc.* **1947**, *69*, 2836–2848. [CrossRef]
39. Lagergren, S. Zur Theorie der Sogenannten Adsorption Gelöster Stoffe, Kungliga Svenska Vetenskapsakademiens. *Handlingar* **1898**, *24*, 1–39.
40. Ho, Y.S.; McKay, G. Pseudo-second order model for sorption processes. *Process Biochem.* **1999**, *34*, 451–465. [CrossRef]
41. Weber, W.J.; Morris, J.C. Intraparticle diffusion during the sorption of surfactants onto activated carbon. *J. Sanit. Eng. Div. Am. Soc. Civ. Eng.* **1963**, *89*, 53–61.
42. Checkol, F.; Elfwing, A.; Greczynski, G.; Mehretie, S.; Inganäs, O.; Admassie, S. Highly Stable and Efficient Lignin-PEDOT/PSS Composites for Removal of Toxic Metals. *Adv. Sustain. Syst.* **2018**, *2*, 1700114. [CrossRef]
43. Karnib, M.; Kabbani, A.; Holail, H.; Olama, Z. Heavy metals removal using activated carbon, silica and silica activated carbon composite. *Energy Procedia* **2014**, *50*, 113–120. [CrossRef]
44. Nyairo, W.N.; Eker, Y.R.; Kowenje, C.; Akin, I.; Bingol, H.; Tor, A.; Ongeri, D.M. Efficient adsorption of lead (II) and copper (II) from aqueous phase using oxidized multiwalled carbon nanotubes/polypyrrole composite. *Sep. Sci. Technol.* **2018**, *53*, 1498–1510. [CrossRef]
45. Badruddoza, A.Z.M.; Shawon, Z.B.Z.; Tay, W.J.D.; Hidajat, K.; Uddin, M.S. Fe_3O_4/cyclodextrin polymer nanocomposites for selective heavy metals removal from industrial wastewater. *Carbohydr. Polym.* **2013**, *91*, 322–332. [CrossRef] [PubMed]
46. Liu, Y.; Liu, Z.; Gao, J.; Dai, J.; Han, J.; Wang, Y.; Xie, J.; Yan, Y. Selective adsorption behavior of Pb (II) by mesoporous silica SBA-15-supported Pb (II)-imprinted polymer based on surface molecularly imprinting technique. *J. Hazard. Mater.* **2011**, *186*, 197–205. [CrossRef] [PubMed]
47. Fallah, Z.; Isfahani, H.N.; Tajbakhsh, M.; Tashakkorian, H.; Amouei, A. TiO_2-grafted cellulose via click reaction: An efficient heavy metal ions bioadsorbent from aqueous solutions. *Cellulose* **2018**, *25*, 639–660. [CrossRef]
48. Khalek, M.D.; Mahmoud, G.A.; El-Kelesh, N.A. Synthesis and Characterization of Poly-Methacrylic Acid Grafted Chitosan-Bentonite Composite and Its Application for Heavy Metals Recovery. *Chem. Mater. Res.* **2012**, *2*, 1–12.
49. Ghoul, M.; Bacquet, M.; Morcellet, M. Uptake of heavy metals from synthetic aqueous solutions using modified PEI—Silica gels. *Water Res.* **2003**, *37*, 729–734. [CrossRef]

Article

Deposition of Boron-Doped Thin CVD Diamond Films from Methane-Triethyl Borate-Hydrogen Gas Mixture

Nikolay Ivanovich Polushin [1], Alexander Ivanovich Laptev [1], Boris Vladimirovich Spitsyn [2], Alexander Evgenievich Alexenko [2], Alexander Mihailovich Polyansky [3], Anatoly Lvovich Maslov [1,*] and Tatiana Vladimirovna Martynova [1]

1 Research Laboratory of Superhard Materials, National University of Science and Technology MISIS, Leninsky avenue, 4, Moscow 119049, Russia; polushin@misis.ru (N.I.P.); laptev@misis.ru (A.I.L.); martynova.t@misis.ru (T.V.M.)

2 Institute of Physical Chemistry, Russian Academy of Sciences, Leninsky avenue, 31, Moscow 119991, Russia; bvspitsyn@gmail.com (B.V.S.); alexenko@inbox.ru (A.E.A.)

3 JSC "NPO Energomash named after Academician V.P. Glushko", Burdenko street, 1, Khimki 141401, Russia; energo@npoem.ru

* Correspondence: almaslov@misis.ru

Received: 30 April 2020; Accepted: 2 June 2020; Published: 4 June 2020

Abstract: Boron-doped diamond is a promising semiconductor material that can be used as a sensor and in power electronics. Currently, researchers have obtained thin boron-doped diamond layers due to low film growth rates (2–10 μm/h), with polycrystalline diamond growth on the front and edge planes of thicker crystals, inhomogeneous properties in the growing crystal's volume, and the presence of different structural defects. One way to reduce structural imperfection is the specification of optimal synthesis conditions, as well as surface etching, to remove diamond polycrystals. Etching can be carried out using various gas compositions, but this operation is conducted with the interruption of the diamond deposition process; therefore, inhomogeneity in the diamond structure appears. The solution to this problem is etching in the process of diamond deposition. To realize this in the present work, we used triethyl borate as a boron-containing substance in the process of boron-doped diamond chemical vapor deposition. Due to the oxygen atoms in the triethyl borate molecule, it became possible to carry out an experiment on simultaneous boron-doped diamond deposition and growing surface etching without the requirement of process interruption for other operations. As a result of the experiments, we obtain highly boron-doped monocrystalline diamond layers with a thickness of about 8 μm and a boron content of 2.9%. Defects in the form of diamond polycrystals were not detected on the surface and around the periphery of the plate.

Keywords: CVD process; doping; single-crystal diamond; boron; triethyl borate; thin films; boron-doped diamond

1. Introduction

Today, there is a significant number of publications devoted to the study of electrophysical properties of boron-doped single-crystal synthetic diamond [1–7]. Other works are devoted to reviews of the mechanism and optimal conditions for thin diamond layer deposition; some of these works are connected with boron-doped diamonds [8–11]. In [12,13], it was shown that the thicker the diamond layer is grown, the higher the probability of defects occurring at its growing surface. The number of defects is notably affected by the single-crystal CVD-diamond (chemical vapor deposition-diamond) deposition rate: the probability of defect formation and their number per unit area grows with the increase of deposition rate. Therefore, boron-doped single-crystal diamond thin films are grown at a low deposition rate. The necessity of intermediate treatment of the obtained samples and loading of new substrates into the reactor chamber leads to the expenditure of large amounts of time and resources. Such production costs lead to higher costs and lower output of the product.

To reduce the cost of production (for example, sensors or crystals for power electronics) and to increase production, growing thick homogeneous boron-doped single-crystal CVD-diamond layers was proposed as rough material for the jewelry industry or for cutting thin doped diamond plates for subsequent treatment.

Reducing the number of defects on the surface (Figure 1) and increasing the boron-doped single-crystal CVD-diamond deposition rate can be achieved by etching the surface of the growing diamond during the deposition process.

Figure 1. The surface of large thickness diamond with defects on the surface and facets (substrate size—5 × 5 × 0.5 mm, deposited diamond layer thickness—0.5 mm).

Apparently, it is necessary to keep the composition of the gas phase constant in order to receive stable properties over the whole volume of the diamond. In addition, controlling the nitrogen content in the diamond is unavoidable, which can be achieved by using high-purity substances. Oxygen could be added into the gas mixture to etch the surface of a growing diamond. In this case, the problem of synthesis of thick boron-doped single-crystal diamond films is narrowed down to solving the following subtasks:

1. To find an appropriate boron-containing substance.
2. To optimize the process of synthesis of boron-doped diamond of small thickness.
3. To optimize the process of synthesis of thick layers of boron-doped diamond due to its constant etching.

After the realization of the third point, it becomes possible not only to reduce the cost of semiconductor products significantly but also to obtain fancy-colored diamonds for the jewelry industry.

Today, researchers use various boron-containing compounds for doping diamond during its deposition from the gas phase; the most typical additives are demonstrated in Table 1.

Table 1. Frequently used doping diamond boron-containing substances.

Substance	Coating Thickness, μm	References
$B(CH_3)_3$	1.5–25	[14–17]
B_2H_6	5–350	[18–21]
BCl_3	0.5–1.0	[22]
$(CH_3O)_3B$	2–>100	[23,24]
$(C_2H_5O)_3B$	–	[25,26]
hBN	770–1020	[27]

As can be seen from Table 1, in most experiments, diborane is used for doping diamond with boron. This compound, under standard conditions, is a colorless gas with a nauseating sweet odor. Diborane is qualified as an unstable flammable substance. Besides this, this gas is highly toxic and poses a hazard to humans [28–32]. This substance is used as an acceptor dopant for epitaxial silicon in the industrial production of semiconductor materials. It is also used for doping silicon plates by the deposition/dispersion technique and polysilicon interconnection of some capacitors. Diborane is widely used in the formation of borosilicate glasses [33,34] and often used as a source of hydrogen in many chemical reactions [35].

Diborane belongs to the class of boranes. These compounds have high chemical reactivity, major heat of combustion, and they are not thermodynamically stable compounds of boron and hydrogen. Many of the boranes, due to the nuances listed above, belong to substances with the highest category of flammability: they can self-ignite from high temperatures not only in air but also in contact with water or a number of halogenated hydrocarbons [36]. Table 2 represents the characteristics of some boranes: their stability and behavior in air and water [37].

Table 2. Some characteristics of the most important boranes [33].

Compound	Thermal Stability	Behavior on Air	Reaction with H_2O
B_2H_6	Stable at 25 °C	Self-ignites	Hydrolyzes immediately
B_4H_{10}	Unstable at 25 °C	Self-ignites in the presence of water	Hydrolyzes
B_5H_9	Stable at 25 °C	Self-ignites	Hydrolyzes by heating
B_5H_{11}	Unstable at 150 °C and above	Self-ignites	Hydrolyzes rapidly
B_6H_{10}	Unstable at 25 °C	Stable	Hydrolyzes by heating
B_6H_{12}	Unstable at 25 °C	Stable	Hydrolyzes by heating
B_9H_{15}	Unstable at 25 °C	Stable	Hydrolyzes by heating
$B_{10}H_{14}$	Stable at 150 °C	Extremely stable	Hydrolyzes slowly

As a result of the analysis of Table 2, it is possible to conclude that boranes, except older boranes (starting with B_6H_{10}), are unstable compounds. At the same time, despite the absence of such a serious drawback, such as autoignition in the presence of oxygen, these boranes are not widely used in both scientific and industrial applications.

Diborane, which is prevalent in the semiconductor industry, is never used with oxygen. Trimethyl borate is more volatile, flammable, and faster hydrolyzed than triethyl borate. Trimethyl borate boils at 68 °C, triethyl borate at 119 °C [38].

Taking into account the properties of boranes, as well as the need to use oxygen to achieve the objectives, we decided to use triethyl borate $(C_2H_5O)_3B$.

2. Materials and Methods

2.1. Materials

In this research, we used an excess of boric acid (99.9% purity) in ethyl alcohol (95% aqueous solution) as initial stock for triethyl borate. During the synthesis, a constant temperature of the mixture was maintained at 25 ± 1 °C. To input triethyl borate to the Ardis 300, we used a hydrogen blow with least 99.9995% purity. The methane purity was not less than 99.995%.

A single-crystal diamond with misorientation angles and directions that corresponded to the requirements for growing single-crystal diamonds was used as a substrate for deposition.

2.2. Scanning Electron Microscopy and X-ray Microanalysis

The study of the surface layers of the coatings was carried out on a HITACHI SU5000 scanning electron microscope equipped with annexes for X-ray spectral microanalysis X-Max^{N} and electron backscatter diffraction (EBSD) Nordlys Nano from OXFORD INSTRUMENTS. Phase identification was accomplished using the NIST Structural Database (NSD).

2.3. The Hardness and Elastic Modulus of Coatings

The hardness and elastic modulus of boron-doped diamond films were determined using the Oliver–Farr method [39,40] on a Micro Indentation Tester CSM-Instruments (Switzerland), using a Vickers diamond pyramid. The maximum load was 50.00 mN.

According to the Oliver–Farr method, hardness is the ratio of the maximum indentation load and the projection area of the indenter contact to the material without elastic deflection at the edge of the print (i.e., without the elastic component).

The formula for hardness is the following [41] (Figure 2):

$$H = \frac{F}{S} = \frac{F}{k \cdot h_c^2}$$

where F—max force, S—contact area,

h_c—depth of contact of the indenter with the material, defined as

$$h_c = h_{max} - \frac{\varepsilon \cdot P_{max}}{S} = h_{max} - \varepsilon \cdot \left(h_{max} - h_i\right)$$

ε *and* k—constants depending on the indenter geometry,

h_i—distance corresponding to the intersection of the tangent to the unloading curve at the beginning part of the displacement axis,

h_{max}—maximum depth of indenter penetration in the material.

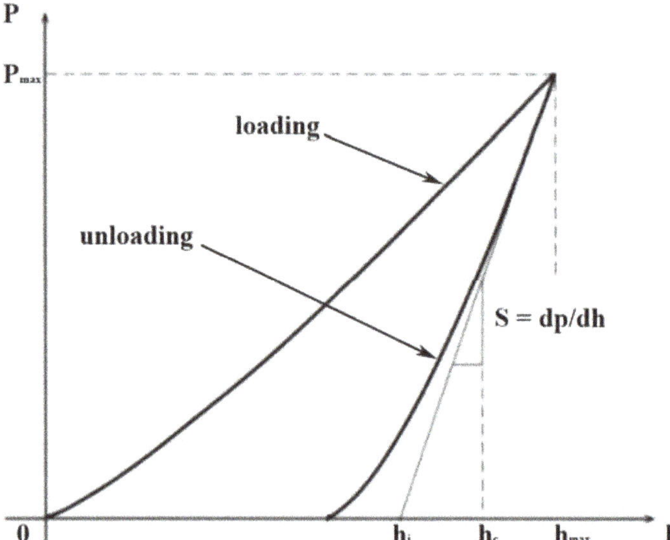

Figure 2. Scheme of loading–unloading curve.

Contact stiffness (S) is determined by the slope of the initial part of the unloading curve (P_{max}):

$$S = \left(\frac{dP}{dh}\right)_{P\,=\,P_{max}}$$

2.4. Preparing the Single-Crystal Diamond Surface and the Thin-Layer Doping Process

Figure 3 shows a schematic diagram of the single-crystal diamond substrate preparation for doping with boron and the subsequent treatment of products.

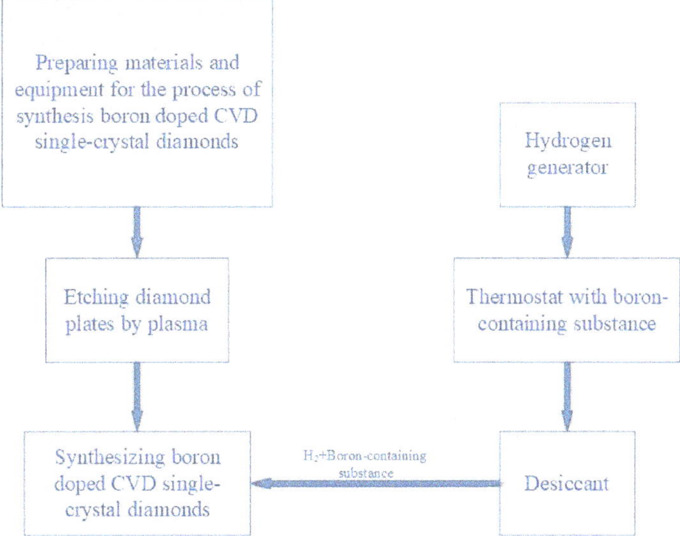

Figure 3. Schematic diagram of the technology of preparing single-crystal diamond substrates for doping and the doping process itself.

The preparation stages for materials and equipment, including the preparation of diamond substrates by etching, laser cutting, and subsequent mechanical treatment, are the same with the single-crystal diamond substrate preparation for high-purity single-crystal diamond deposition by the CVD process.

The process of substrates etching can be carried out in various atmospheres. For example, diamond is etched at a low rate in hydrogen plasma [42]; in oxygen-argon mixture, it is much higher and reaches 200 nm/min [43]; in an oxygen-hydrogen mixture (with up to 2% oxygen), the etching rate is more than 200 nm/min [44].

The doped single-crystal films were grown from the gas phase, which comprised methane, hydrogen, and a boron-containing substance, on single-crystal diamond substrates. To achieve this, as previously shown, trimethyl or triethyl borate can be used. These substances are obtained by mixing boric acid and alcohol—either methyl or ethyl. In this work, we used ethanol to prepare triethyl borate $(C_2H_5O)_3B$. Electrolytically generated hydrogen after dehydration and purification was divided into two streams. A smaller part (5–15 cm^3/min) was passed through a bubbler system with ethanol and boric acid that formed volatile triethyl borate. The second part of the hydrogen from the generator was input into the reactor. In order to keep the composition of the gas mixture unchanged during the whole process of deposition, we used a saturated solution of boric acid. Boric acid in the bubbler system was always maintained in excess. The bubbler system was placed in a thermostat at a temperature of 25 °C. Silica gel desiccant was used to dry the input mixture. The desiccant was a steel cylinder with a diameter of 40 mm and a length of 300 mm and it was filled with silica gel. For ease of exploitation (drying silica gel after work), the cylinder was equipped with a quick-detachable connection with a soldered metal mesh to exclude silica gel contact with the system.

Hydrocarbon radicals with high reactivity were formed in the gas phase as a result of the activation of the mixture of hydrogen, methane, and triethyl borate. As a result of decomposing hydrocarbon radicals on the substrate, free carbon and atomic hydrogen were released, which ensures the high selectivity of the diamond growth process due to the release of nondiamond carbon that could crystallize on the substrate [45].

The deposition process was carried out at 9.806 kPa pressure in the system. The substrate temperature was 1100 °C. The methane flow to the reactor was 25 cm^3/min; the flow of a mixture of triethyl borate and hydrogen (with possible vapors of ethanol) was 10 cm^3/min.

The required parameters of pressure and gas flow were maintained automatically. The duration of the deposition process was 2 h at a growth rate of 4 µm/h.

3. Results and Discussions

3.1. Scanning Electron Microscopy and Elemental Analysis

Figure 4 shows the samples obtained as a result of the experiments.

Figure 4. Boron-doped diamond samples (Sample 1—left, Sample 2—right).

As seen in Figure 4, both samples have surfaces without such defects as on the example of the plate in Figure 1. A large amount of triethyl borate was supplied to the system with the purpose of increasing the oxygen concentration in the plasma. Obviously, such high concentrations of oxygen allowed etching defects. Figure 5 demonstrates the surface images of Samples 1 and 2 with boron-doped layers.

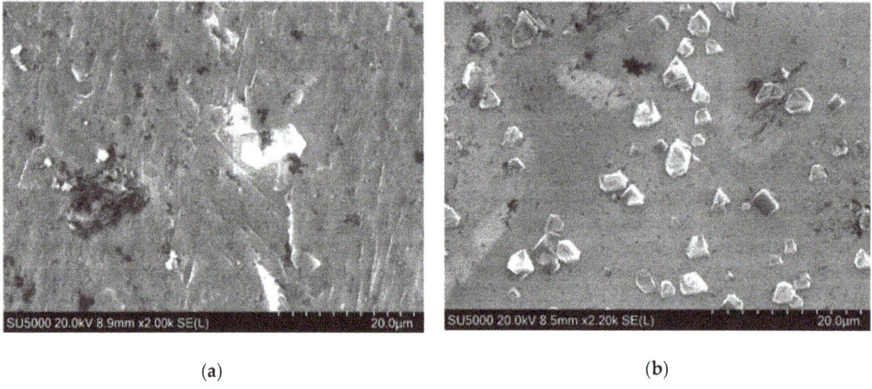

(**a**) (**b**)

Figure 5. SEM images of the sample surfaces. (**a**) Sample 1, (**b**) Sample 2.

It can be seen that the surface of Sample 1 has growth "waves". In addition, there are incoherent formations on it. The surface of Sample 2 also contains incoherent formations, but the growth "waves" are less noticeable.

The distribution of Sample 1 areas for determining the chemical composition using X-ray microanalysis is presented in Figure 6. Table 3 presents the results of Sample 1 surface X-ray microanalysis.

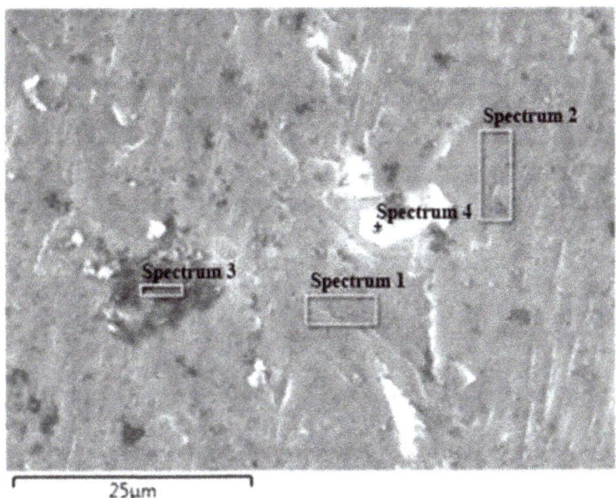

Figure 6. SEM image of Sample 1 with areas for identifying surface composition.

Table 3. Sample 1 elemental analysis.

The Number of Area	C	B	O
1	97.1	2.9	–
2	97.2	2.8	–
3	92.2	3.2	4.6
4	94.1	3.4	2.5

Table 3 (data from Figure 7), containing elemental analysis based on the EDS (energy-dispersive X-ray spectroscopy) results, demonstrates that the first position, as well as the second, corresponds to heavily boron-doped diamond with a boron content of 2.9%.

The third position is a foreign substance containing carbon, oxygen, and boron that corresponds to the composition of triethyl borate, which, apparently, could depose from the gas phase at the end of the diamond-doping process.

The distribution of Sample 2 areas for determining the chemical composition using X-ray microanalysis is presented in Figure 8. On the surface of Sample 2, as well as on Sample 1, there are a significant number of foreign formations.

Table 4 (data from Figure 9) contains elemental analysis based on the EDS and presents the results of Sample 2 surface X-ray microanalysis. The boron content in Sample 2 was 2.9%, as in Sample 1. Considering that samples were obtained in independent experiments with the same synthesis parameters, the result indicates high doping-process stability.

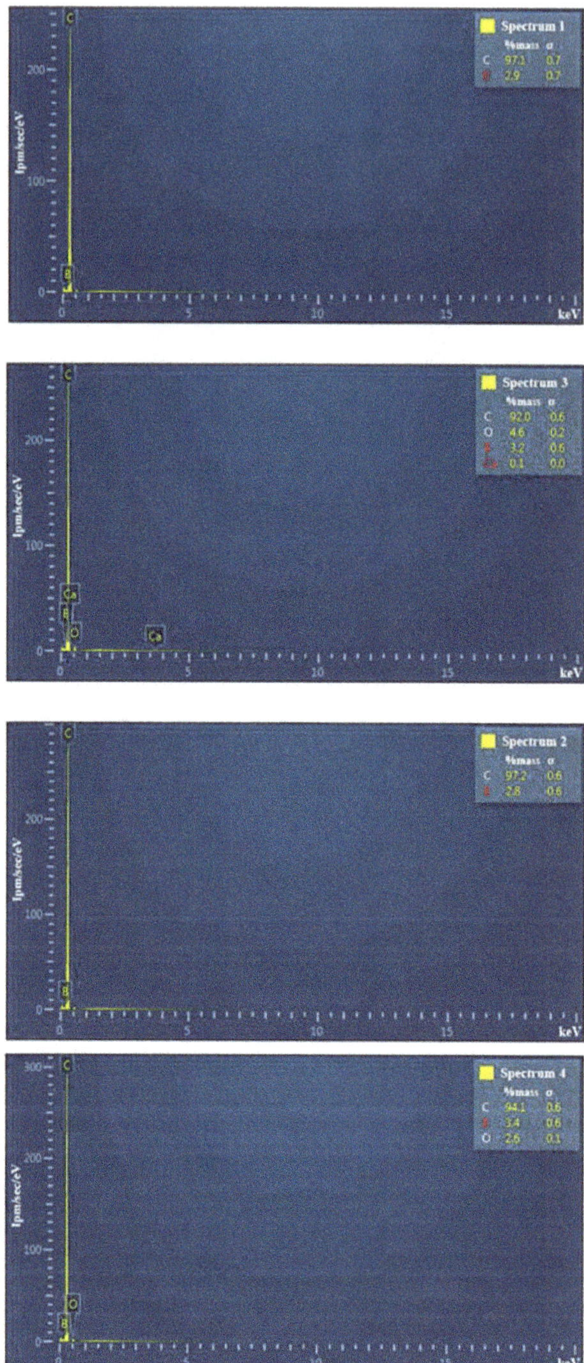

Figure 7. EDS analysis of sample 1.

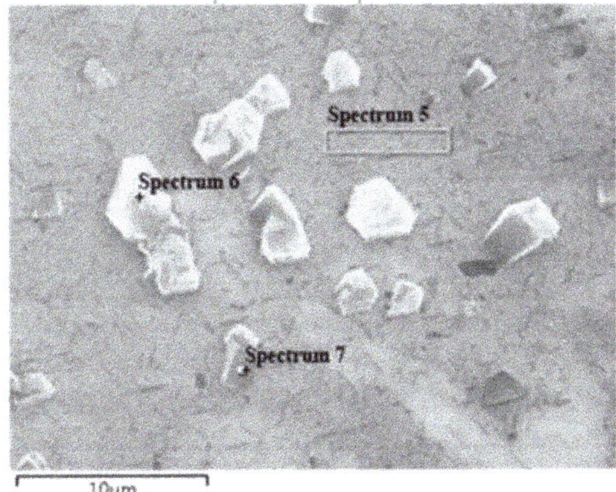

Figure 8. Areas for identifying Sample 2 surface composition.

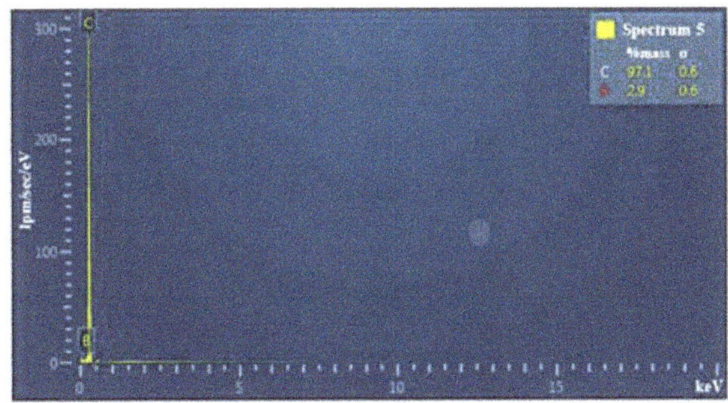

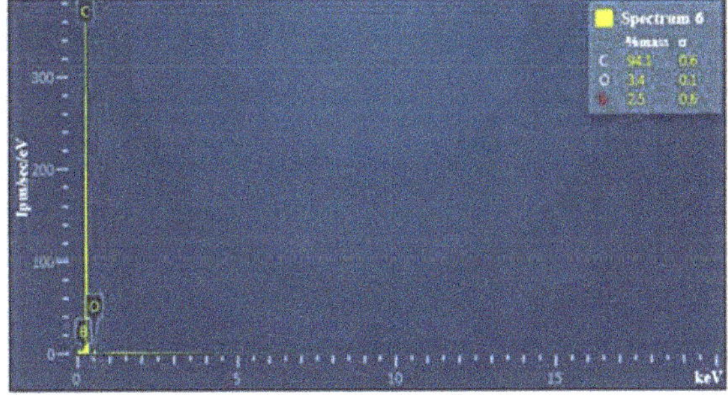

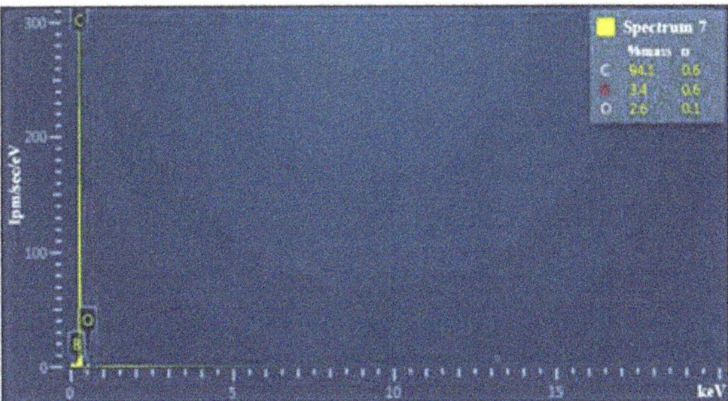

Figure 9. EDS analysis of Sample 2.

Table 4. Sample 2 elemental analysis.

The Number of Area	C	B	O
5	97.1	2.9	–
6	94.1	2.5	3.4
7	94.1	3.4	2.5

Positions 7 and 8 are a foreign substance corresponding to the composition of triethyl borate, which, apparently, could depose from the gas phase at the end of the diamond-doping process. To eliminate such defects in both cases, it will be sufficient to etch the obtained diamond surface in the stream of atomic hydrogen without doping or carbon-containing substances.

Using the EBSD method, we obtained electron diffraction patterns, indexed them, determined their phase composition, type of lattice, and orientation (Euler angles) in each of the six investigated areas. We revealed that the coating orientations in different areas were the same, which allowed us to conclude that the coating was single-crystal and was formed by epitaxial growth.

3.2. The Hardness and Elastic Modulus

We obtained values of hardness in the range from 62 to 117 GPa and elastic modulus in the range from 914 to 1099 GPa. A typical loading–unloading curve is represented in Figure 10.

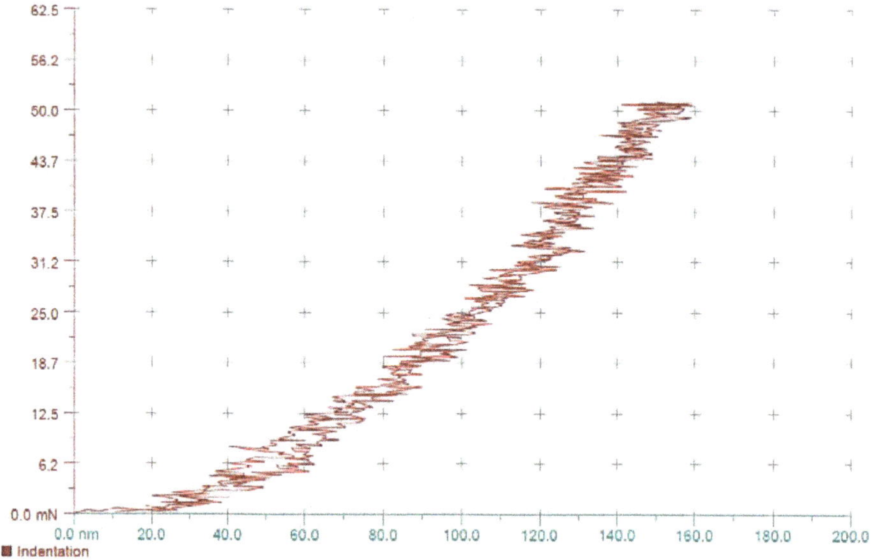

Figure 10. Loading–unloading curve.

From the presented load–unloading curve, we identified the following characteristics:

- hardness 62.8 GPa;
- elastic modulus 914.6 GPa;
- indenter penetration depth 155.816 nm.

We qualified the optimal mode for the synthesis of heavily boron-doped diamond:

- temperature 1100 °C;
- H_2 flow rate 480 cm^3/min;

- CH_4 flow rate 25 cm^3/min;
- $H_2 - (C_2H_5O)_3B$ flow rate 10 cm^3/min;
- microwave power 3800 W;
- reactor pressure 9.806 kPa.

The obtained experimental data allow further experiments to reduce the boron concentration in single-crystal diamond film and to introduce more oxygen to etch the surface during the deposition process of thick homogeneous boron-doped diamond layers.

4. Conclusions

The paper presents a scheme of boron-doped diamond CVD synthesis by using triethyl borate as a boron-containing substance. The replacement of the boron-containing substance from diborane to triethyl borate allowed etching, creating polycrystalline diamond via an all-synthesis process without the need for interruption due to oxygen atoms from triethyl borate. The high boron content obtained according to the proposed scheme samples (2.9% mass) is explained by the need for high oxygen atoms concentration in the gas mixture for successful etching of the emerging polycrystalline diamond. As a result of synthesis, it is possible to obtain boron-doped diamond with a stable boron distribution in the volume.

Using the Oliver–Farr method, we measured the hardness and elastic modulus of obtained coatings, varying from 62 to 117 and 914 to 1099 GPa, respectively.

In the future, work will be developed in the direction of reducing the concentration of triethyl borate in the gas mixture and optimizing parameters for sectoral growth in order to obtain doped diamond layers. Compensation of oxygen atoms will be realized through direct oxygen supply to the gas mixture, which is stable in contrast to diborane.

Author Contributions: Conceptualization, N.I.P. and B.V.S.; methodology, A.I.L., A.E.A., and A.M.P.; validation, A.E.A., A.L.M., and A.M.P.; formal analysis, A.I.L. and A.L.M.; investigation, N.I.P., B.V.S., A.E.A., and T.V.M.; resources, N.I.P.; data curation, N.I.P.; writing—original draft preparation, A.L.M. and T.V.M.; writing—review and editing, A.L.M. and T.V.M.; visualization, T.V.M.; supervision, B.V.S.; project administration, N.I.P.; funding acquisition, N.I.P. All authors have read and agreed to the published version of the manuscript.

Funding: This research was funded by the Ministry of Education and Science of the Russian Federation, grant number 075-15-2019-255, dated 9 July 2019, a unique identifier of the agreement RFMEFI57818X0266.

Conflicts of Interest: The authors declare no conflict of interest.

References

1. Jia, F.; Bai, Y.; Qu, F.; Zhao, J.-J.; Zhuang, C.; Jiang, X. Effect of B/C ratio on the physical properties of highly boron-doped diamond films. *Vacuum* **2010**, *84*, 930–934. [CrossRef]
2. Rehacek, V.; Hotový, I.; Marton, M.; Mikolasek, M.; Michniak, P.; Vincze, A.; Kromka, A.; Vojs, M. Voltammetric characterization of boron-doped diamond electrodes for electroanalytical applications. *J. Electroanal. Chem.* **2020**, *862*, 114020. [CrossRef]
3. Zhang, J.; Yu, X.; Zhang, Z.-Q.; Zhao, Z.-Y. Preparation of boron-doped diamond foam film for supercapacitor applications. *Appl. Surf. Sci.* **2020**, *506*, 144645. [CrossRef]
4. Tarelkin, S.; Bormashov, V.; Pavlov, S.; Kamenskyi, D.; Kuznetsov, M.; Terentiev, S.; Prikhodko, D.; Galkin, A.; Hubers, H.-W.; Blank, V. Evidence of linear Zeeman effect for infrared intracenter transitions in boron doped diamond in high magnetic fields. *Diam. Relat. Mater.* **2017**, *75*, 52–57. [CrossRef]
5. Ficek, M.; Sobaszek, M.; Gnyba, M.; Ryl, J.; Gołuński, Ł.; Śmietana, M.; Jasiński, J.; Caban, P.; Bogdanowicz, R. Optical and electrical properties of boron doped diamond thin conductive films deposited on fused silica glass substrates. *Appl. Surf. Sci.* **2016**, *387*, 846–856. [CrossRef]
6. Bormashov, V.; Tarelkin, S.; Buga, S.; Kuznetsov, M.; Terentiev, S.; Semenov, A.; Blank, V. Electrical properties of the high quality boron-doped synthetic single-crystal diamonds grown by the temperature gradient method. *Diam. Relat. Mater.* **2013**, *35*, 19–23. [CrossRef]

7. Mortet, V.; Pernot, J.; Jomard, F.; Soltani, A.; Remes, Z.; Barjon, J.; D'Haen, J.; Haenen, K. Properties of boron-doped epitaxial diamond layers grown on (110) oriented single crystal substrates. *Diam. Relat. Mater.* **2015**, *53*, 29–34. [CrossRef]
8. Bolshakov, A.; Zyablyuk, K.; Kolyubin, V.; Dravin, V.; Khmelnitskii, R.; Nedosekin, P.; Pashentsev, V.; Tyurin, E.; Ralchenko, V. Thin CVD diamond film detector for slow neutrons with buried graphitic electrode. *Nucl. Instrum. Methods Phys. Res. Sect. A* **2017**, *871*, 142–147. [CrossRef]
9. Halliwell, S.; May, P.; Fox, N.; Othman, M. Investigations of the co-doping of boron and lithium into CVD diamond thin films. *Diam. Relat. Mater.* **2017**, *76*, 115–122. [CrossRef]
10. Ohmagari, S.; Srimongkon, K.; Yamada, H.; Umezawa, H.; Tsubouchi, N.; Chayahara, A.; Shikata, S.; Mokuno, Y. Low resistivity p+ diamond (100) films fabricated by hot-filament chemical vapor deposition. *Diam. Relat. Mater.* **2015**, *58*, 110–114. [CrossRef]
11. Demlow, S.N.; Rechenberg, R.; Grotjohn, T.; Nicley, S.S. The effect of substrate temperature and growth rate on the doping efficiency of single crystal boron doped diamond. *Diam. Relat. Mater.* **2014**, *49*, 19–24. [CrossRef]
12. Tallaire, A.; Achard, J.; Silva, F.; Sussmann, R.S.; Gicquel, A.; Rzepka, E. Oxygen plasma pre-treatments for high quality homoepitaxial CVD diamond deposition. *Phys. Status Solidi* **2004**, *201*, 2419–2424. [CrossRef]
13. Chen, C.-F.; Chen, S.-H.; Hong, T.-M.; Wang, T.-C. Boron-doped diamond films using trimethylborate as a dopant source in methane-carbon dioxide gas mixtures. *Diam. Relat. Mater.* **1994**, *3*, 632–637. [CrossRef]
14. Okushi, H. High quality homoepitaxial CVD diamond for electronic devices. *Diam. Relat. Mater.* **2001**, *10*, 281–288. [CrossRef]
15. Arima, K.; Miyatake, H.; Teraji, T.; Ito, T. Effects of vicinal angles from (001) surface on the Boron-doping features of high-quality homoepitaxial diamond films grown by the high-power microwave plasma chemical-vapor-deposition method. *J. Cryst. Growth* **2007**, *309*, 145–152. [CrossRef]
16. Teraji, T.; Wada, H.; Yamamoto, M.; Arima, K.; Ito, T. Highly efficient doping of boron into high-quality homoepitaxial diamond films. *Diam. Relat. Mater.* **2006**, *15*, 602–606. [CrossRef]
17. Volpe, P.-N.; Arnault, J.-C.; Tranchant, N.; Chicot, G.; Pernot, J.; Jomard, F.; Bergonzo, P. Boron incorporation issues in diamond when TMB is used as precursor: Toward extreme doping levels. *Diam. Relat. Mater.* **2012**, *22*, 136–141. [CrossRef]
18. Issaoui, R.; Achard, J.; Silva, F.; Tallaire, A.; Tardieu, A.; Gicquel, A.; Pinault, M.A.; Jomard, F. Growth of thick heavily boron-doped diamond single crystals: Effect of microwave power density. *Appl. Phys. Lett.* **2010**, *97*, 182101. [CrossRef]
19. De Feudis, M.; Mille, V.; Valentin, A.; Brinza, O.; Tallaire, A.; Tardieu, A.; Issaoui, R.; Achard, J. Ohmic graphite-metal contacts on oxygen-terminated lightly boron-doped CVD monocrystalline diamond. *Diam. Relat. Mater.* **2019**, *92*, 18–24. [CrossRef]
20. Tallaire, A.; Valentin, A.; Mille, V.; William, L.; Pinault-Thaury, M.; Jomard, F.; Barjon, J.; Achard, J. Growth of thick and heavily boron-doped (113)-oriented CVD diamond films. *Diam. Relat. Mater.* **2016**, *66*, 61–66. [CrossRef]
21. Barjon, J.; Chikoidze, E.; Jomard, F.; Dumont, Y.; Pinault-Thaury, M.-A.; Issaoui, R.; Brinza, O.; Achard, J.; Silva, F. Homoepitaxial boron-doped diamond with very low compensation. *Phys. Status Solidi* **2012**, *209*, 1750–1753. [CrossRef]
22. Dorsch, O.; Holzner, K.; Werner, M.; Obermeier, E.; Harper, R.; Johnston, C.; Chalker, P.R.; Buckley-Golder, I. Piezoresistive effect of boron-doped diamond thin films. *Diam. Relat. Mater.* **1993**, *2*, 1096–1099. [CrossRef]
23. Zubkov, V.; Kucherova, O.V.; Bogdanov, S.A.; Zubkova, A.V.; Butler, J.; Ilyin, V.; Afanas'Ev, A.V.; Vikharev, A.L. Temperature admittance spectroscopy of boron doped chemical vapor deposition diamond. *J. Appl. Phys.* **2015**, *118*, 145703. [CrossRef]
24. Bogdanov, S.; Vikharev, A.; Drozdov, M.; Radishev, D. Synthesis of thick and high-quality homoepitaxial diamond with high boron doping level: Oxygen effect. *Diam. Relat. Mater.* **2017**, *74*, 59–64. [CrossRef]
25. Jiang, X.; Willich, P.; Paul, M.; Klages, C.-P. In situ boron doping of chemical-vapor-deposited diamond films. *J. Mater. Res.* **1999**, *14*, 3211–3220. [CrossRef]
26. Srikanth, V.V.S.S.; Kumar, P.S.; Kumar, V.B. A Brief Review on the In Situ Synthesis of Boron-Doped Diamond Thin Films. *Int. J. Electrochem.* **2012**, *2012*, 218393. [CrossRef]
27. Yap, C.; Ansari, K.; Xiao, S.; Yee, S.; Chukka, R.; Misra, D. Properties of near-colourless lightly boron doped CVD diamond. *Diam. Relat. Mater.* **2018**, *88*, 118–122. [CrossRef]

28. National Library of Medicine National Center for Biotechnology Information. Boranes. Available online: https://pubchem.ncbi.nlm.nih.gov/compound/Diborane (accessed on 29 April 2020).
29. National Research Council. *Acute Exposure Guideline Levels for Selected Airborne Chemicals*; National Academies Press: Washington, DC, USA, 2003; Volume 3, p. 4.
30. Pan, L.; Lew, K.-K.; Redwing, J.M.; Dickey, E.C. Effect of diborane on the microstructure of boron-doped silicon nanowires. *J. Cryst. Growth* **2005**, *277*, 428–436. [CrossRef]
31. Gandía, J.; Cárabe, J.; Swinnen, J. A simple diborane-degradation model for controlling p-type doping of microcrystalline silicon. *Sol. Energy Mater. Sol. Cells* **2002**, *73*, 75–90. [CrossRef]
32. Rojwal, V.; Singha, M.K.; Mondal, T.; Mondal, D. Formation of micro structured doped and undoped hydrogenated silicon thin films. *Superlattices Microstruct.* **2018**, *124*, 201–217. [CrossRef]
33. Kim, C.-H.; Jung, S.-H.; Jeon, J.-H.; Han, M.-K. A simple low-temperature laser-doping employing phosphosilicate glass and borosilicate glass films for the source and drain formation in poly-Si thin film transistors. *Thin Solid Films* **2001**, *397*, 4–7. [CrossRef]
34. Lottspeich, F.; Müller, M.; Köhler, R.; Fischer, G.; Schneiderlöchner, E.; Palinginis, P. Influence of nitride and nitridation on the doping properties of PECVD-deposited BSG layers. *Energy Procedia* **2017**, *124*, 691–699. [CrossRef]
35. Mikhailov, B.M. The Chemistry of Diborane. *Russ. Chem. Rev.* **1962**, *31*, 207–224. [CrossRef]
36. Kauffman, G.B. ENCYCLOPÆDIA BRITANNICA. Borane. Available online: https://www.britannica.com/science/borane (accessed on 29 April 2020).
37. Cotton, F.A.; Wilkinson, G.; Gaus, P.L. *Chemia Nieorganiczna*; Wydawnictwo Naukowe: Warszawa, Poland, 1995.
38. Docks, E.L. *Boric Acid Esters*; Wiley: Hoboken, NJ, USA, 2000.
39. Ast, J.; Ghidelli, M.; Durst, K.; Göken, M.; Sebastiani, M.; Korsunsky, A.M.; Goeken, M. A review of experimental approaches to fracture toughness evaluation at the micro-scale. *Mater. Des.* **2019**, *173*, 107762. [CrossRef]
40. Ghidelli, M.; Sebastiani, M.; Collet, C.; Guillemet, R. Determination of the elastic moduli and residual stresses of freestanding Au-TiW bilayer thin films by nanoindentation. *Mater. Des.* **2016**, *106*, 436–445. [CrossRef]
41. Moshchenok, V.I.; Lyakhovitsky, M.M.; Doshechkina, I.V.; Kukhareva, I.E. Comparison of the calculation method for assessing the surface nano- and microhardness of materials with the Oliver-Farr method. *Metals* **2009**, *8*, 719.
42. Man, W.; Wang, J.; Wang, C.; Wang, S.; Xiong, L. Planarizing CVD diamond films by using hydrogen plasma etching enhanced carbon diffusion process. *Diam. Relat. Mater.* **2007**, *16*, 1455–1458. [CrossRef]
43. Enlund, J.; Isberg, J.; Karlsson, M.; Nikolajeff, F.; Olsson, J.; Twitchen, D.J. Anisotropic dry etching of boron doped single crystal CVD diamond. *Carbon* **2005**, *43*, 1839–1842. [CrossRef]
44. Bolshakov, A.P.; Ralchenko, V.G.; Polsky, A.V.; Ashkinazi, E.E.; Homich, A.A.; Sharanov, G.V.; Hmelnitsky, R.A.; Zavedeev, E.V.; Homich, A.V.; Sovik, D.N. Synthys single crystal diamond in microwave plasma. *Prikladnaya Fizika* **2011**, *6*, 104–110.
45. Spitsyn, B.V.; Aleksenko, A.E. Chemical crystallization of diamond and deposition diamond coating from the gas phase. *Met. Prot.* **2007**, *5*, 456–474. [CrossRef]

Communication

Selective Photocatalytic Oxidation of 5-HMF in Water over Electrochemically Synthesized TiO_2 Nanoparticles

Anna Ulyankina [1,*], Sergey Mitchenko [2] and Nina Smirnova [1]

1 Platov South-Russian State Polytechnic University (NPI), 346428 Novocherkassk, Russia; smirnova_nv@mail.com
2 Institute of Physical-Organic Chemistry and Coal Chemistry, 83114 Donetsk, Ukraine; samit_rpt@mail.ru
* Correspondence: anya-barbashova@yandex.ru

Received: 2 May 2020; Accepted: 28 May 2020; Published: 29 May 2020

Abstract: TiO_2 nanoparticles were prepared via an electrochemical method using pulse alternating current and applied in the photocatalytic oxidation of 5-hydroxymethylfurfural (HMF) to 2,5-diformylfuran (DFF). Its physicochemical properties were characterized by SEM, HRTEM, XRD, and BET methods. The effect of scavenger and UVA light intensity was studied. The results revealed that electrochemically synthesized TiO_2 nanoparticles exhibit higher DFF selectivity in the presence of methanol (up to 33%) compared with commercial samples.

Keywords: 5-hydroxymethylfurfural; 2,5-diformylfuran; TiO_2; electrochemical synthesis; pulse alternating current

1. Introduction

5-hydroxymethylfurfural (HMF) is one of the most versatile "biomass platform molecules" produced from abundant and renewable lignocellulose-derived glucose. HMF can be further converted into a series of high-value chemicals such as 2,5-diformylfuran (DFF), 5-hydroxymethyl-2-furancarboxylic acid (HMFCA), and 2,5-furandicarboxylic acid (FDCA) [1]. Among them, DFF is regarded as a valuable starting material for the polymer, pharmaceutical, and other industries [2]. The catalytic (both homogeneous and heterogeneous) oxidation of HMF is the most promising and extensively studied method to produce DFF. However, these catalytic processes are often carried out in organic solvents and at elevated temperatures and pressures, which makes them energy intensive and harmful to the environment [3,4]. Heterogeneous photocatalysis has become an attractive technology for air and wastewater treatment due to the easy in situ formation of powerful oxidizing agents in the presence of photocatalysts under mild conditions (ambient temperature and pressure, solar radiation, etc.). [5,6]. Recently, a lot of efforts have been made in the field of selective and partial photocatalytic oxidation or reduction of diverse substrates to form compounds of higher value in aqueous solutions (organic synthesis) [7,8]. However, to control the selectivity of HMF oxidation to DFF in an aqueous medium is challenging. It is reported that a number of factors, such as the type of photocatalyst and crystallinity, as well as the type of irradiation, can affect the photocatalytic oxidation of HMF to synthesize DFF [9,10]. A lot of research is devoted to the application of new promising photoactive materials [11]. However, TiO_2-based photocatalysts are more attractive for use in photocatalytic technology due to their stability (both chemical and physical) and nontoxicity [12,13]. It is well documented in the literature that physicochemical properties, such as phase and surface structure, light-absorbing ability, and morphology, vary greatly depending on the synthesis method, and even a small change in the preparation procedure can lead to a large difference in photocatalytic activity [13]. Electrochemical methods are a green and beneficial alternative to traditional synthetic methods. They are easy to scale and very attractive for industrial applications.

In our previous studies, the pulse alternating current (PAC) synthesis was described as a promising strategy producing highly active nanosized materials based on transition metal oxides, including SnO_2–SnO nanocomposite [14], ZnO [15], Co_3O_4/CoOOH [16], and CuO_x [17] for various applications.

In the present study, TiO_2 nanoparticles (TiO_2-NPs), which have been prepared electrochemically under pulsed alternating current conditions, were used in the oxidation of HMF as a photocatalyst.

2. Materials and Methods

2.1. Photocatalyst Preparation

TiO_2-NPs were synthesized by an electrochemical process using pulse alternating current (PAC) followed by thermal treatment. An aqueous electrolyte solution was prepared by dissolving a NaCl salt in distilled water at a concentration of 2 M. An electrochemical cell equipped with a cooling jacket was filled with the electrolyte solution. The Ti plates (99.7%) used as electrodes were immersed into the cell parallel to each other at a distance of 1 cm between them. The asymmetric PAC was then adjusted using a self-made source. The anodic:cathodic average current density ratio (j_a:j_k) was 1.0:0.2 A/cm^2. The current waveform used in the study is illustrated in Figure 1.

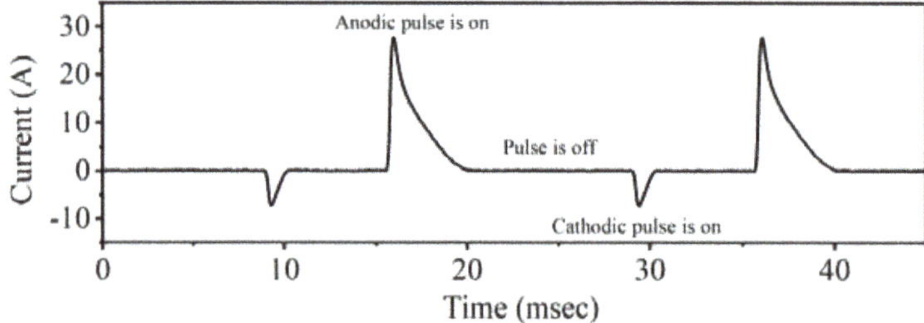

Figure 1. The current waveform during pulse alternating current (PAC) synthesis.

A magnetic stirrer inside the cell maintained the mixing speed in the solution at 400 rpm. A reaction time of 20 min was selected to prevent a considerable deterioration of the Ti electrode. After synthesis, the product was separated by filtration and washed with distilled water. The collected sample was dried in an oven at 80 °C, and then calcined at 500 °C in air for 3 h.

2.2. Characterization

The crystalline phase was analyzed by X-ray diffraction (XRD) on an ARL X'TRA diffractometer using Cu Kα radiation (λ = 1.5406 Å) in 0.02° increments. The size of nanoparticles was calculated using the Scherrer equation:

$$d_{XRD} = \frac{K\lambda}{\beta cos\theta} \quad (1)$$

where d_{XRD} is crystallite size (nm), λ, θ, and β are X-ray wavelength, Bragg diffraction angle, and full width at half-maximum (FWHM) of the diffraction peak, respectively.

Morphological characteristics were observed using a scanning electron microscope (SEM) Tescan Vega 3 SB with Oxford Instruments X-Act EDS detector. Fine structural analysis was performed on a JEOL JEM-2100 transmission electron microscope (TEM). The main lattice spacing of the crystallite structure was calculated by means of a fast Fourier transform analysis of high-resolution TEM images. Specific surface area analysis via the Brunauer–Emmett–Teller (BET) method was done on a Quantachrome NOVA 1200e instrument at 77.3 K.

2.3. Photocatalytic Procedure

Photocatalytic reactions were carried out in a quartz beaker (100 mL) under ambient conditions. The suspension was irradiated with a spotlight source (Hamamatsu, LC8) at a wavelength of 365 nm. The suspension in the beaker was irradiated from the top; the radiation energy ranged from 3.0 to 6.0 mW cm^{-2}, which was measured using a UV34 Lux Meter (PCE). The initial concentration of the used substrate and photocatalyst was 0.4 mM and 1.0 g L^{-1}, respectively. The experiments were carried out under aerobic conditions with continuous bubbling of O_2 at a flow rate of 3 mL/min. A run was also carried out with the addition of methanol and Na_2CO_3 as the scavenger for hole and OH•, respectively. Before turning on the lamp, the suspension was kept in an ultrasonic bath for 5 min, and then stirred for 30 min to achieve adsorption–desorption equilibrium. The amount of HMF adsorbed by the catalyst in the dark was fairly low, i.e., less than 2% of the initial HMF amount. During the run, the aliquots of the aqueous suspension were taken at fixed time intervals and centrifuged (15,000 rpm) before analysis.

The reaction solution was analyzed by high-performance liquid chromatography (HPLC) instrument (Agilent 1260 LC, stable bond C18 column). The mobile phase solution, consisting of 70% acetonitrile and 30% ultrapure water, was supplied to the column at 30 °C and a flow rate of 0.5 mL min^{-1}. The amounts of HMF and DFF were determined from a calibration curve obtained using standard solutions of these compounds. The following formula was used to calculate the selectivity to DFF:

$$Select.(\%) = \frac{M_{t,DFF}}{M_{o,HMF} - M_{t,HMF}} \times 100\% \quad (2)$$

where $M_{t,HMF}$ and $M_{t,DFF}$ are the quantities of moles at a moment t of HMF and DFF, respectively, and the $M_{o,HMF}$ is the initial amount of HMF.

3. Results and Discussion

Figure 2a shows the XRD pattern of the obtained TiO_2. The main diffraction peaks of (101), (103), (004), (112), (200), (105), (211), (213), (204), (116), (220), and (215) correspond to the crystal planes of anatase phase. It is well known that anatase TiO_2 exhibits more appropriate photocatalytic properties compared to rutile TiO_2 [18]. Using Equation (1), the value of crystallite size of TiO_2 NPs was determined to be 17.4 nm. The calculation was in good agreement with the TEM results.

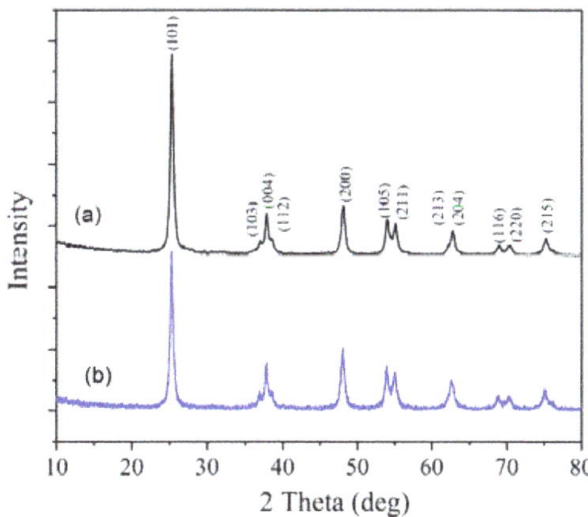

Figure 2. XRD patterns of TiO_2 nanoparticles before (**a**) and after (**b**) photocatalysis.

The EDS spectrum (inset in Figure 3a) shows the presence of peaks associated with the elements Ti (0.4 keV and 4.5 keV) and O (0.5 keV), confirming the characteristic components of TiO_2.

The HRTEM image (Figure 3b) shows individual nanocrystals with clear one-dimensional lattice fringes. The interplanar distance of 0.35 nm of the TiO_2 NPs correspond to the (101) plane of anatase [19].

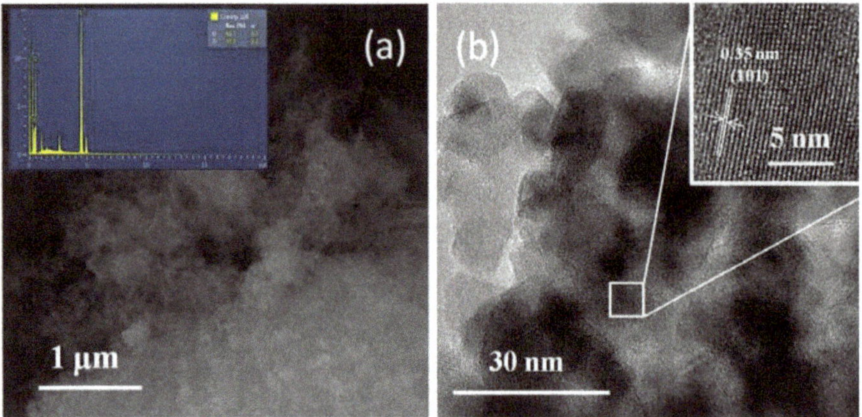

Figure 3. SEM (**a**) and HRTEM (**b**) images of TiO_2 nanoparticles.

The specific surface area of TiO_2 NPs synthesized by the electrochemical method using pulse alternating current was measured using BET analysis. Figure 4 displays the N_2 adsorption–desorption isotherms of TiO_2. The sample presents a type-IV isotherm with a H3 type hysteresis. The BET surface area was calculated to be 58.6 $m^2 g^{-1}$.

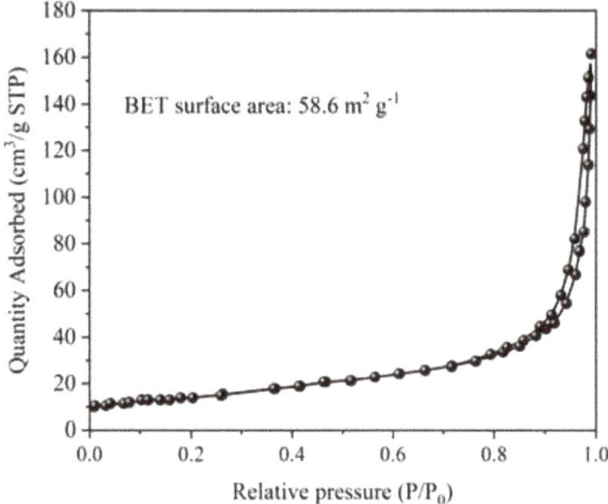

Figure 4. N_2 adsorption–desorption isotherm of TiO_2 nanoparticles.

The detected product of HMF photooxidation is mainly DFF, which is consistent with literature [20]. In the dark or when exposed to light in the absence of a catalyst, no DFF was obtained, suggesting that both light and catalyst are necessary for this photocatalytic oxidation reaction. The efficiency of the electrochemically synthesized TiO_2 NPs in the HMF photooxidation reaction with and without scavengers was evaluated (Figure 5a). The corresponding DFF selectivity values were demonstrated in Figure 5b. As can be seen, HMF can be almost completely oxidized without adding any trapping species over 360 min at a light intensity of 3 mW cm^{-2} with DFF selectivity of approximately 10%. Moreover, HMF degradation during self-photolysis cannot be ignored. It should be considered, however, that the contribution of photolysis in the presence of photocatalyst is always less significant due to the shielding effect of the suspended powder.

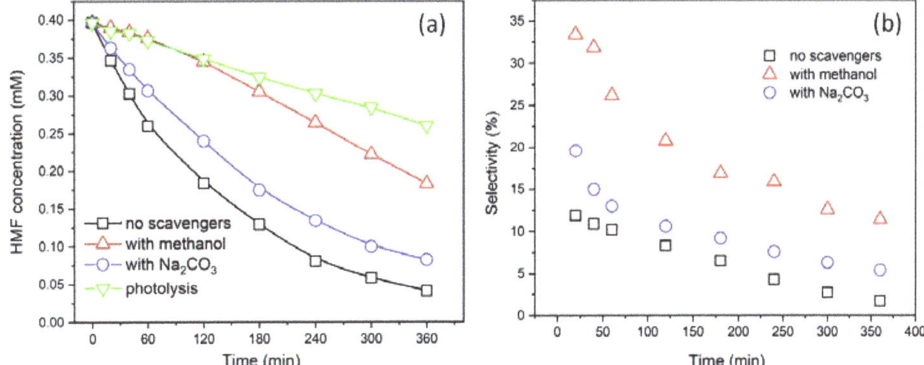

Figure 5. Effect of scavengers on 5-hydroxymethylfurfural (HMF) photocatalytic oxidation (**a**) and 2,5-diformylfuran (DFF) selectivity (**b**). Experimental conditions: light irradiation intensity 3.0 mW cm^{-2}, catalyst dosage 1 g L^{-1}, HMF concentration 0.4 mM, O_2 flow 3 mL min^{-1}.

In order to improve the selectivity for the partial oxidation reaction, methanol and Na_2CO_3 were added to the reaction suspension as a hole and an OH• scavenger, respectively. It was observed that a strong decrease in HMF oxidation efficiency was accompanied by an increase in the selectivity of DFF compared to a similar run carried out in the absence of methanol. It is noteworthy that the Na_2CO_3 trapping agent had a slight effect on DFF selectivity during HMF oxidation. These results indicate that in a methanol-containing photocatalytic system, alcohol acts as an efficient hole trap by preventing the deep mineralization of HMF and increasing selectivity for the target DFF. In a previous paper [9], a similar trend was observed for poorly crystallized TiO_2. The evolution of DFF slowed down with time (Figure 5b), which may be due to its overoxidation.

The effect of UVA light intensity on DFF selectivity was also examined. It was found that an increase in light intensity leads to an increase in overall efficiency of HMF degradation (Figure 6a), but decreases the selectivity of DFF (Figure 6b), probably due to higher contribution of direct photolysis processes at higher light intensities [11].

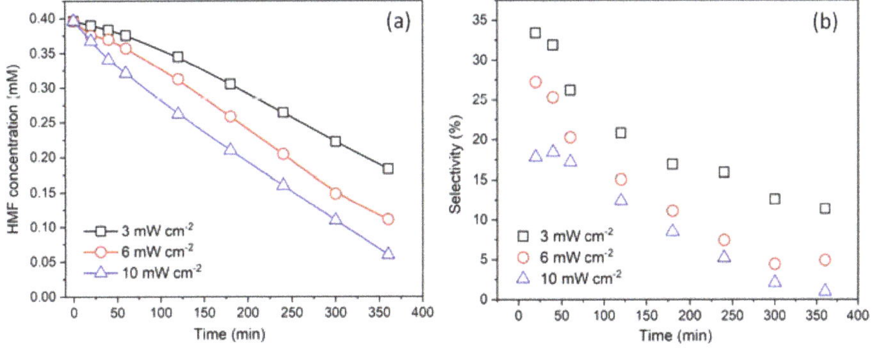

Figure 6. Experimental results on photocatalytic oxidation of HMF (**a**) and DFF selectivity (**b**) depending on irradiation intensity in the presence of electrochemically synthesized TiO_2 and methanol. Experimental conditions: catalyst dosage 1 g L^{-1}, HMF concentration 0.4 mM, methanol concentration 50 mM, O_2 flow 3 mL min^{-1}.

The obtained selectivity toward DFF (up to 33%) for electrochemically synthesized TiO_2 NPs was higher than for the previously reported titanium oxide samples prepared by microemulsion and sol–gel methods [9,12]. A run carried out for comparison with the commercial TiO_2 Evonik P25 (not reported here) showed that the commercial catalyst was more active in the decomposition of HMF; however, in this case, DFF was formed only in trace amounts (~1%). This may be due to the higher level of crystallinity in the commercial sample [9]. A comparison between the photocatalytic performance of the electrochemically synthesized TiO_2 NPs and other TiO_2 photocatalysts for HMF oxidation is presented in Table 1.

Table 1. Comparison of photocatalytic performance of the TiO_2 nanoparticles (NPs) with other TiO_2 photocatalysts for HMF oxidation.

Sample	HMF Concentration	Irradiation Type	HMF Conversion (%)	DFF Selectivity (%)	Reference
TiO_2-m	0.08 M	300 W Xe-lamp (250–2500 nm)	10 12 22	2 2 5	[12]
HPA [1]	0.5 mM	four 8 W Philips lamps (365 nm) 3.0 mW cm^{-2}	20	26	[9]
TiO_2 NPs [1]	0.4 mM	Hamamatsu, LC8 (365 nm) 3.0 mW cm^{-2}	15	29.5	the present study
comm-TiO_2	0.4 mM	Hamamatsu, LC8 (365 nm) 3.0 mW cm^{-2}	60	1.0	the present study

[1]—Run performed in the presence of 50 mM methanol.

The reusability of the synthesized TiO_2 NPs has been verified by recycling experiments during which TiO_2 NPs was reused by centrifugation and washing with water. Recycling experiments suggest that HMF could be converted into DFF in three consecutive runs using recollected photocatalyst. It is worth mentioning that the selectivity values became even higher after the second run (Figure 7) than can be attributed to the decrease in TiO_2 crystallinity.

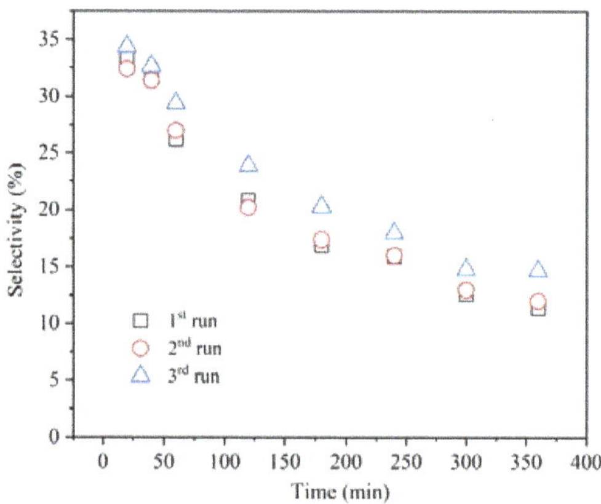

Figure 7. Cycling runs in photocatalytic oxidation of HMF over electrochemically synthesized TiO_2.

The stability of the TiO_2 photocatalyst after the HMF photocatalytic oxidation has also been investigated using XRD analysis, which is shown in Figure 2b. The results suggested that the crystallinity of the TiO_2 NPs becomes slightly lower after its third usage, which may be due to the photocorrosion of the catalyst. In addition, the higher background in the diffraction pattern of the TiO_2 sample after photocatalysis is consistent with the appearance of some amorphous phase in the sample. However, no other diffraction peaks were observed after the third photocatalytic run.

4. Conclusions

Photocatalytic oxidation of HMF in water can provide an environmentally benign method to produce valuable DFF. Electrochemical synthesis using pulse alternating current was carried out to prepare TiO_2 nanoparticles with a higher selectivity toward DFF (up to 33%) compared to the commercial TiO_2 sample. Photocatalytic selective oxidation of HMF to form DFF using electrochemically synthesized TiO_2 can be scaled up after optimization of all parameters.

Author Contributions: Investigation, A.U.; writing—original draft preparation, A.U.; writing—review and editing, S.M.; supervision, N.S. All authors have read and agreed to the published version of the manuscript.

Funding: This research was funded by RUSSIAN SCIENCE FOUNDATION, grant number 16-13-10444.

Acknowledgments: The authors thank the Shared Research Center "Nanotechnologies" of Platov South-Russian State Polytechnic University (NPI) for HPLC and XRD analyses.

Conflicts of Interest: The authors declare no conflict of interest.

References

1. Klushin, V.A.; Galkin, K.I.; Kashparova, V.P.; Krivodaeva, E.A.; Kravchenko, O.A.; Smirnova, N.V.; Chernyshev, V.M.; Ananikov, V.P. Technological aspects of fructose conversion to high-purity 5-hydroxymethylfurfural, a versatile platform chemical. *Russ. J. Org. Chem.* **2016**, *52*, 767–771. [CrossRef]
2. Ghosh, K.; Molla, R.A.; Iqubal, M.A.; Islam, S.S.; Islam, S.M. Ruthenium nanoparticles supported on N-containing mesoporous polymer catalyzed aerobic oxidation of biomass-derived 5-hydroxymethylfurfural (HMF) to 2,5-diformylfuran (DFF). *Appl. Catal. A Gen.* **2016**, *520*, 44–52. [CrossRef]
3. Teng, N.; Li, J.-L.; Lu, B.-Q.; Wang, Y.-Q.; Jia, S.-Y.; Wang, Y.-X.; Hou, X.-L. The selective aerobic oxidation of 5-hydroxymethylfurfural to produce 2,5-diformylfuran using Nitrogen-doped porous carbons as catalysts. *New Carbon Mater.* **2019**, *34*, 593–599. [CrossRef]
4. Ding, L.; Yang, W.; Chen, L.; Cheng, H.; Qi, Z. Fabrication of spinel $CoMn_2O_4$ hollow spheres for highly selective aerobic oxidation of 5-hydroxymethylfurfural to 2,5-diformylfuran. *Catal. Today* **2018**. [CrossRef]
5. Mamaghani, A.H.; Haghighat, F.; Lee, C.-S. Hydrothermal/solvothermal synthesis and treatment of TiO_2 for photocatalytic degradation of air pollutants: Preparation, characterization, properties, and performance. *Chemosphere* **2019**, *219*, 804–825. [CrossRef] [PubMed]
6. Wang, R.; Ma, X.; Liu, T.; Li, Y.; Song, L.; Tjong, S.C.; Cao, L.; Wang, W.; Yu, Q.; Wang, Z. Degradation aspects of endocrine disrupting chemicals: A review on photocatalytic processes and photocatalysts. *Appl. Catal. A Gen.* **2020**, *597*, 117547. [CrossRef]
7. Akbari, A.; Amini, M.; Tarassoli, A.; Eftekhari-Sis, B.; Ghasemian, N.; Jabbari, E. Transition metal oxide nanoparticles as efficient catalysts in oxidation reactions. *Nano Struct. Nano Objects* **2018**, *14*, 19–48. [CrossRef]
8. Chen, L.; Tang, J.; Song, L.-N.; Chen, P.; He, J., Au, C.-T., Yin, S.-F. Heterogeneous photocatalysis for selective oxidation of alcohols and hydrocarbons. *Appl. Catal. B Environ.* **2019**, *242*, 379–388. [CrossRef]
9. Yurdakal, S.; Tek, B.S.; Alagöz, O.; Augugliaro, V.; Loddo, V.; Palmisano, G.; Palmisano, L. Photocatalytic Selective Oxidation of 5-(Hydroxymethyl)-2-furaldehyde to 2,5-Furandicarbaldehyde in Water by Using Anatase, Rutile, and Brookite TiO_2 Nanoparticles. *ACS Sustain. Chem. Eng.* **2013**, *1*, 456–461. [CrossRef]
10. Wu, Q.; He, Y.; Zhang, H.; Feng, Z.; Wu, Y.; Wu, T. Photocatalytic selective oxidation of biomass-derived 5-hydroxymethylfurfural to 2,5-diformylfuran on metal-free g-C_3N_4 under visible light irradiation. *Mol. Catal.* **2017**, *436*, 10–18. [CrossRef]
11. Zhang, H.; Feng, Z.; Zhu, Y.; Wu, Y.; Wu, T. Photocatalytic selective oxidation of biomass-derived 5-hydroxymethylfurfural to 2,5-diformylfuran on WO_3/g-C_3N_4 composite under irradiation of visible light. *J. Photochem. Photobiol. A Chem.* **2019**, *371*, 1–9. [CrossRef]
12. Lolli, A.; Maslova, V.; Bonincontro, D.; Basile, F.; Ortelli, S.; Albonetti, S. Selective Oxidation of HMF via Catalytic and Photocatalytic Processes Using Metal-Supported Catalysts. *Molecules* **2018**, *23*, 2792. [CrossRef] [PubMed]
13. Lozano-Morales, S.A.; Morales, G.; López Zavala, M.Á.; Arce-Sarria, A.; Machuca-Martínez, F. Photocatalytic Treatment of Paracetamol Using TiO_2 Nanotubes: Effect of pH. *Processes* **2019**, *7*, 319. [CrossRef]

14. Ulyankina, A.A.; Kuriganova, A.B.; Smirnova, N.V. Photocatalytic properties of SnO_2–SnO nanocomposite prepared via pulse alternating current synthesis. *Mendeleev Commun.* **2019**, *29*, 215–217. [CrossRef]
15. Ulyankina, A.; Leontyev, I.; Avramenko, M.; Zhigunov, D.; Smirnova, N. Large-scale synthesis of ZnO nanostructures by pulse electrochemical method and their photocatalytic properties. *Mater. Sci. Semicond. Process.* **2018**, *76*, 7–13. [CrossRef]
16. Chernysheva, D.; Vlaic, C.; Leontyev, I.; Pudova, L.; Ivanov, S.; Avramenko, M.; Allix, M.; Rakhmatullin, A.; Maslova, O.; Bund, A.; et al. Synthesis of Co_3O_4/CoOOH via electrochemical dispersion using a pulse alternating current method for lithium-ion batteries and supercapacitors. *Solid State Sci.* **2018**, *86*, 53–59. [CrossRef]
17. Ulyankina, A.; Leontyev, I.; Maslova, O.; Allix, M.; Rakhmatullin, A.; Nevzorova, N.; Valeev, R.; Yalovega, G.; Smirnova, N. Copper oxides for energy storage application: Novel pulse alternating current synthesis. *Mater. Sci. Semicond. Process.* **2018**, *73*, 111–116. [CrossRef]
18. Luttrell, T.; Halpegamage, S.; Tao, J.; Kramer, A.; Sutter, E.; Batzill, M. Why is anatase a better photocatalyst than rutile?—Model studies on epitaxial TiO_2 films. *Sci. Rep.* **2014**, *4*, 4043. [CrossRef] [PubMed]
19. Sathiyan, K.; Bar-Ziv, R.; Mendelson, O.; Zidki, T. Controllable synthesis of TiO_2 nanoparticles and their photocatalytic activity in dye degradation. *Mater. Res. Bull.* **2020**, *126*, 110842. [CrossRef]
20. Ma, B.; Wang, Y.; Guo, X.; Tong, X.; Liu, C.; Wang, Y.; Guo, X. Photocatalytic synthesis of 2,5-diformylfuran from 5-hydroxymethyfurfural or fructose over bimetallic Au-Ru nanoparticles supported on reduced graphene oxides. *Appl. Catal. A Gen.* **2018**, *552*, 70–76. [CrossRef]

Article

Effect of the Chemical Composition on the Structural State and Mechanical Properties of Complex Microalloyed Steels of the Ferritic Class

Alexander Zaitsev [1,2], Anton Koldaev [1], Nataliya Arutyunyan [1,2,*], Sergey Dunaev [2] and Dmitrii D'yakonov [1]

1 Bardin Central Research Institute of Ferrous Metallurgy, Moscow 105005, Russia; aizaitsev1@yandex.ru (A.Z.); koldaevanton@gmail.com (A.K.); aberkas@yandex.ru (D.D.)
2 Faculty of Chemistry, Lomonosov Moscow State University, Moscow 119991, Russia; dunaev@general.chem.msu.ru
* Correspondence: naarutyunyan@gmail.com; Tel.: +7-495-939-1673

Received: 28 April 2020; Accepted: 27 May 2020; Published: 29 May 2020

Abstract: The most promising direction for obtaining a unique combination of difficult-to-combine properties of low-carbon steels is the formation of a dispersed ferrite microstructure and a volumetric system of nanoscale phase precipitates. This study was aimed at establishing the special features of the composition influence on the characteristics of the microstructure, phase precipitates, and mechanical properties of hot-rolled steels of the ferritic class. It was carried out by transmission electron microscopy and testing the mechanical properties of metal using 8 laboratory melts of low-carbon steels microalloyed by V, Nb, Ti, and Mo in various combinations. It was found that block ferrite prevails in the structure of steel cooled after hot rolling at a rate of 10–15 °C/s. Lowering of the microalloying components content leads to a decrease in the block ferrite fraction to 20–35% and the dominance of polygonal ferrite. The presence of nanoscale carbide (carbonitride) precipitates of austenitic and interphase/mixed types was detected in the rolled steels. It was established that the tendencies of changes in the characteristics of the structural state and present phase precipitates correlate well with obtained values of strength properties. The advantages of titanium-based microalloying systems in comparison with vanadium-based are shown.

Keywords: low-carbon steels of the ferritic class; nanoscale phase precipitates; strength characteristics; steel composition; hot rolling; structure

---

1. Introduction

Currently, research and development of new types of structural steels is preferably aimed at achieving simultaneously high indicators of difficult-to-combine properties: strength, ductility, formability, corrosion resistance, weldability, and other service characteristics while reducing production costs [1]. This is due to an increase in requirements for critical parts used in various branches of engineering and industry: construction, engineering, transport, mining, oil, and others. The use of traditional methods for increasing the strength of steel, which are based on the implementation of such hardening mechanisms as grain boundary, solid solution, and dispersion hardening, causes a deterioration in ductility and other service properties. The search for fundamentally new solutions has led to the development of advanced steels for automobile sheet: dual-phase ferrite-martensitic and multiphase steels with complex phase composition, as well as with increased ductility achieved as a result of effects of transformation- and twinning-induced plasticity [1,2]. However, such steels, due to the presence of hard components (martensite, bainite, etc.) in their structure, have limited formability, and their production requires the use of specialized equipment and a complex alloying system, leading to increased costs [2]. Thus, to date, there are no steels combining simultaneously high indicators of strength, ductility, formability, fatigue and corrosion resistance, and other service properties.

The most promising direction for solving the formulated problem is the creation of new steels with a homogeneous ductile ferrite microstructure strengthened by formation of a volumetric system of nanoscale phase precipitates and ferrite grain refinement [3–9]. Both strengthening mechanisms, in many respects, are controlled by phase precipitations of various types and dispersion. The precipitates formed in austenite (austenitic) contribute to the refinement of the microstructure. The dispersion hardening mechanism is realized mainly due to the precipitation of carbides and carbonitrides during or after the $\gamma \rightarrow \alpha$ phase transformation when steel is cooling or tempering [3,10,11]. Such precipitates are named interphase and ferritic, respectively. Interphase precipitates, as a rule, have a size of 1–5 nm. They are arranged in layers with a period of 10–50 nm, parallel to the moving front of the γ/α transformation. In ferrite, disordered formation of nanoscale carbide (carbonitride) precipitates occurs. Their nucleation and growth takes place at dislocations; the average size, as a rule, is 3–10 nm [10].

Initially, to follow these principles, the Ti-Mo microalloying system was used, which provided producing rolled products with a strength of up to 700–1000 MPa and high ductility, formability, and other properties [4,12–14]. It was established that the formation of interphase precipitates, which make a greater contribution to the strengthening than ferritic, occurs at a low (10–15 °C/s) cooling rate of steel after rolling, leading to the simultaneous formation of a polygonal ferrite structure with reduced strength characteristics [4]. Obtaining a stronger high-dislocation structure of block (acicular) ferrite is the result of a higher (~30 °C/s) metal cooling rate, causing phase precipitation in ferrite. Maximum strength indicators can be achieved by creating conditions for the simultaneous formation of an acicular ferrite structure and a system of nanoscale interphase precipitates [4,5].

The results of subsequent studies (e.g., [6,15]) showed the possibility of using more complex V-Nb-Ti-Mo microalloying systems. It served as the basis for the development of technologies for the production of hot-rolled XPF650, XPF800, and XPF1000 steels of 650–1000 MPa strength classes at TATA steel (UK, Netherlands) [7]. However, they are characterized by a high content of manganese and microalloying components (up to wt.%): Mn-2, V-0.32, Nb-0.08, Mo-0.50, and the technological parameters of production are mainly established by trial and error. This is due to the lack of reliable data on the tendencies of the effect of the composition on the characteristics of forming phase precipitates, the structural state, and mechanical properties of complex microalloyed steels of the ferritic class. The present study aims to establish these influence patterns.

2. Materials and Methods

The investigation was carried out for low-carbon steels of the ferritic class, microalloyed by V, Nb, Ti, and Mo in various combinations close to XPF grade. The steels were smelted in a vacuum induction furnace with a magnesite crucible with a steel capacity of 7–8 kg and casted in a heated mold in one ingot. The chemical composition (Table 1) of the investigated steels was obtained by spectral analysis using an emission spectrometer.

Table 1. The chemical composition of the studied steels, wt.%.

No. Steel	C	Si	Mn	P	S	Mo	Al	Ti	Nb	V	N
1	0.059	0.135	1.23	0.002	0.004	0.005	0.036	0.070	0.110	-	0.0070
2	0.053	0.110	1.28	0.003	0.004	-	0.020	0.078	0.030	-	0.0075
3	0.059	0.150	1.36	0.002	0.004	-	0.031	0.110	0.027	-	0.0115
4	0.061	0.150	1.38	0.003	0.004	0.200	0.020	0.170	0.024	-	0.0097
5	0.053	0.110	1.50	0.003	0.005	0.210	0.011	0.069	0.010	0.005	0.0091
6	0.050	0.087	1.47	0.003	0.005	0.190	0.017	0.068	0.010	0.200	0.0135
7	0.048	0.093	1.55	0.003	0.003	0.190	0.016	0.100	0.015	0.010	0.0076
8	0.055	0.088	1.60	0.003	0.006	0.200	0.019	0.130	0.016	0.059	0.0083

It is seen from Table 1 that the produced steels are characterized by a relatively high manganese content of 1.2–1.6 wt.%, as well as by a complex system of separate or joint microalloying of V, Nb, Ti, and Mo with a variation in the concentrations of the components over a relatively wide range. Due to the special feature of laboratory smelting, the produced steels have a relatively high nitrogen content and low phosphorus concentration. As will be shown below, this imposes a certain physical reserve on the established mechanical properties of rolled products due to the negative influence of nitrogen and neutral effect of phosphorus [16]. The steel ingots of all compositions were heated to 1250 °C and rolled on a DUO-300 reversible rolling mill into strips of 3 mm thick with a rolling end temperature of 900 ± 15 °C. After the rolling end, the strip was cooled in an air stream with a cooling rate of 10–15 °C/s to 650 °C. This temperature was chosen based on the previously obtained evidence [17] that to increase the strength properties, it is necessary to use relatively high temperatures for coiling the strip. Then, the strip was placed in a furnace heated to 650 °C, kept for 30 min, followed by cooling with the furnace, simulating the cooling of a roll. Samples were made from the obtained hot-rolled strips.

The microstructure was studied by scanning electron microscopy, SEM, using a JSM-6610LV (JEOL) device equipped with an INCA Energy Feature XT energy dispersive microanalysis system, INCA Wave 500 wave dispersion spectrometer, and transmission electron microscopy, TEM, using a JEM200CX device. Due to big difference between dislocations density in different ferrite types, it was determined by counting dislocations on TEM images related to each type. The type of nanoscale precipitates was identified using the original technique based on analysis of their reflexes on microdiffraction images and a comparison of several characteristics, including the following. Austenitic precipitates always have a tangential scatter of reflexes on microdiffraction images, which means disorientation of the particles, sometimes quite substantial (several degrees). Reflexes from ferritic particles look like long radial bands. Interphase precipitates usually have superposition of both special characteristics of reflexes. The common distinguishing feature of interphase precipitates is their arrangement in layers parallel to the moving front of the γ/α transformation. Particles formed in austenite are characterized by the same orientation of particles in neighboring ferrite grains with different orientations. Another indicator is the Baker–Nutting orientation relationship for precipitates and the ferrite matrix. This relation occurs for ferritic particles, for austenitic—it is not fulfilled, and interphase precipitates have only one orientation relationship [18].

The mechanical properties (yield strength, ultimate tensile strength, and relative elongation) were determined using a HECKERT FP-100/1 tensile testing machine.

The thermodynamic analysis of the regions of phases existence in the studied steels was carried out using the thermodynamic computer model [19] implemented on the basis of proprietary software, by finding the conditions of thermodynamic equilibrium in multicomponent, multiphase systems with given external and internal parameters (temperature, pressure, chemical composition). As a result, the types of equilibrium phases, their quantities and compositions were determined by analysis of the complete possible set for an alloying system of steel under consideration.

3. Results

The results of determining the mechanical properties of the samples of investigated rolled steels are presented in Table 2.

Table 2. Mechanical properties of the studied rolled samples.

No. Steel	Yield Strength, MPa	Ultimate Tensile Strength, MPa	Relative Elongation, %
1	615	680	15
2	535	600	18
3	585	655	16
4	650	765	15
5	555	625	15
6	610	700	14
7	605	665	16
8	600	700	12

As can be seen from Table 2, the strength characteristics of the rolled samples vary over a wide range, rising with increasing in V, Nb, and Ti concentrations and when Mo is present in the steel composition. The values of relative elongation have close values in the absence of a definite correlation with the strength characteristics. To clarify the noted features of the behavior of mechanical properties, a study was carried out of structural state characteristics and present phase precipitates in rolled steels.

By means of SEM, it was found that all studied samples have a close ferrite microstructure. To establish the mechanisms of the influence of chemical composition on the structure formation and properties of the steels under investigation, a detailed study of their structural state and the carbide (carbonitride) precipitates present was made by TEM methods. The results are summarized in Table 3. In all cases, similar tendencies have been established for rolled steels containing Nb, Ti, and for steels of V, Nb, Ti, and Mo complex microalloying. It was found that the metal matrix of rolled samples from steels of all melts consists of ferrite of two morphological types: block and polygonal. Block ferrite, BF, (Figure 1) is characterized by a dislocation density ranging from medium to high. The shape of the blocks is close to equiaxed. The elongated shape of the blocks in the rolled steels No. 1, 2, 3 was not found, in samples of steels containing molybdenum, except steel No. 6—was rarely observed. In rolled steel of smelting No. 6, the shape of ferrite blocks is predominantly elongated and rarely close to equiaxial.

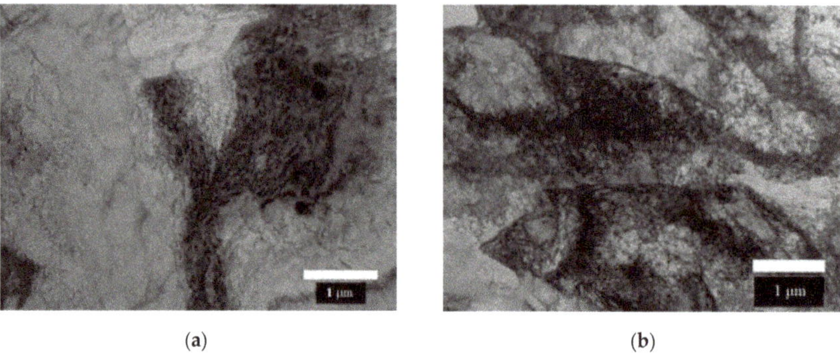

Figure 1. Typical views of block ferrite in rolled steel of smelting: (**a**)—No. 1, (**b**)—No. 6, ×15,000.

Polygonal ferrite, PF, (Figure 2a) has dislocation density varying from low to medium, grain size, as a rule—up to 10–15 μm. In some cases (melts No. 2, 3, 5), there are separate grains larger than 15 μm. In all steels, except No. 2 and 3, the volume fraction of block ferrite is predominant. In rolled steel No. 1, rarely located regions of quasipolygonal ferrite are also detected (Figure 2b) with developed blocking and dislocation density, which is intermediate between that characteristics for PF and BF.

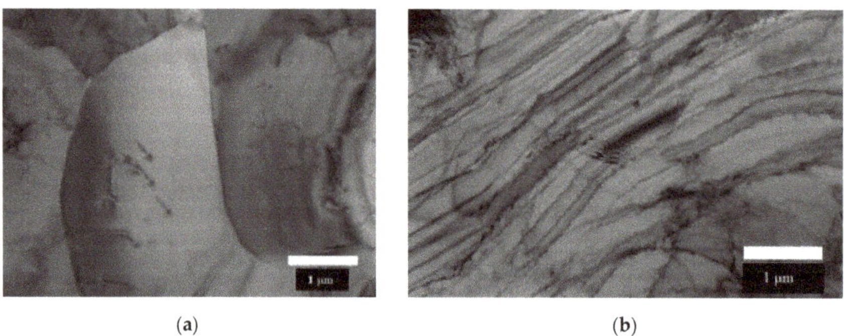

Figure 2. Typical views of polygonal (**a**) and quasipolygonal (**b**) ferrite in rolled steel of smelting No. 1, ×15,000.

Table 3. A summary of the results of the microstructure and phase precipitates investigation of the studied steels.

No. Steel	Characteristics of Microstructure						Characteristics of Precipitates			
	Block Ferrite			Polygonal Ferrite		Cementite, Amount/size	Austenitic		Interphase	
	Fraction, %	Form	Dislocations Density, cm^{-2}	Dislocations Density, cm^{-2}	Grain Size, μm		Grain Fraction, %	Size, nm Length/ Width	Grain Amount/ Amount in a Grain	Size, nm
1	65–75	Equiaxed	$\sim 6 \times 10^{10}$	$\sim 3 \times 10^{9}$	Up to 10–15	Few/ several μm	20	10/up to 3–4	In almost all grains/ significant	Up to 3–4, rarely up to 5–6
2	25–35	Equiaxed	$\sim 2 \times 10^{10}$	$\sim 3 \times 10^{9}$	10–15 there are >15	Few/ several μm	15–20	10/3–4	In almost all grains/less than in steel No 3	Up to 3–4, rarely up to 5–6
3	20–30	Equiaxed	$\sim 2 \times 10^{10}$	$\sim 3 \times 10^{9}$	10–15 there are >15	Few/ several μm	15–20	10/3–4	In almost all grains/less than in steel No 1	Up to 3–4, rarely up to 5–6
4	60–80	Equiaxed, rarely -elongated	$\sim 6 \times 10^{10}$	$\sim 3 \times 10^{9}$	Up to 10–15	No	50	Up to 15/ up to 3–4	In all grains/ significant	Up to 2–4
5	80–85	Equiaxed, rarely -elongated	$\sim 2 \times 10^{10}$	$\sim 3 \times 10^{9}$	10–15 there are >15	Few/ several μm	10	Up to 10/ up to 3–4	Rarely/a few	Up to 3–4
6	90	Elongated	$\sim 2 \times 10^{10}$	$\sim 1 \times 10^{9}$	Up to 10	No	75	Up to 15/ up to 3–4	In all grains/ significant	Up to 3–4
7	80–85	Equiaxed, rarely -elongated	$\sim 6 \times 10^{10}$	$\sim 3 \times 10^{9}$	Up to 10–15	Few/BF: <0.2–0.3μm, PF: up to 1μm	50	Up to 10/ up to 3–4	In all grains/ significant	Up to 2–4
8	85–90	Equiaxed, rarely -elongated	$\sim 6 \times 10^{10}$	$\sim 1 \times 10^{9}$	Up to 10	No	60	Up to 10/ up to 3–4	In all grains/ significant	Up to 2–4

By means of TEM study of thin foils, the presence of austenite in the rolled structure of all the studied steels was not detected. In all melts, with the exception of No. 4, 6, 8, a small amount of cementite was detected. It has the form of separate precipitates along the boundaries of grains/blocks with sizes up to several micrometers (Figure 3). Precipitates of smaller sizes ~0.1–0.2 μm are practically absent. Cementite precipitates were not found in rolled steel from smelting No. 6, and in the case of smelting No. 7, their size in BF was, as a rule, no more than 0.2–0.3 μm, and in PF—up to ~1 μm.

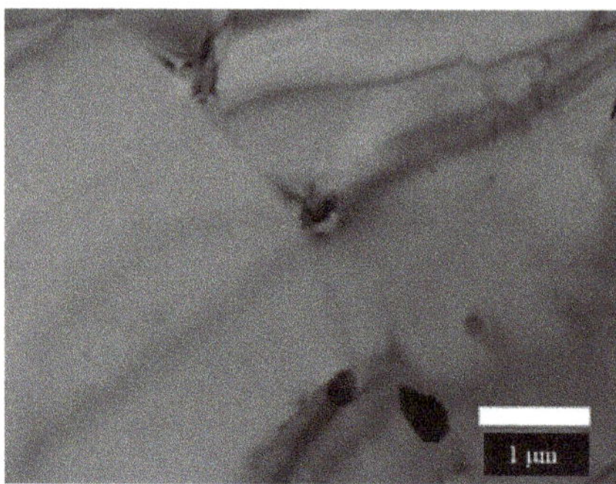

Figure 3. Typical view of cementite precipitate in rolled steel of smelting No. 2, ×15,000.

Detected nanoscale carbide, carbonitride precipitates belong to two dimensional types. The first type includes elongated austenitic precipitates up to ~10 nm in length (in the case of smelting No. 4, 6—up to ~15 nm) and not more than 3–4 nm in width (Figure 4a) formed in austenite. In the case of steels No. 1, 2, 3, they were searched out in approximately 15–20% of ferrite grains, which have a favorable orientation for their detection. Herewith, in those grains/blocks, in which they are present, their number is usually small. In the case of smelting No. 4, 6, 7, and 8, austenitic precipitates are significantly more representative and are found out in about 50–75% of ferrite grains, which have a favorable orientation for their detection. The minimum amount of austenitic precipitates was observed in steel No. 5.

Another size group consists of nanoscale precipitates of interphase/mixed type, which are systematically present both in grains/blocks, in which austenitic precipitates are present, and in grains/blocks, in which they are absent (Figure 4b). Except for steel No. 5, apparently, interphase precipitates are localized in almost all ferrite grains/blocks. The size of precipitates observed in most sections is up to 3–4 nm. Regions where interphase precipitates are larger—up to 5–6 nm are rarely found. For interphase precipitates, the reflexes are noticeably blurred in the tangential direction. Reflexes of complex shape correspond to nanoscale precipitates of a mixed type. Diffraction patterns, in which reflexes from nanoscale precipitates would have the shape of a radially elongated strand (i.e., precipitates would form mainly in ferrite), were not detected.

The results of TEM analysis of carbide precipitates amounts can be explained using thermodynamic calculation of the temperature dependences of the equilibrium phase composition of studied steels. They are presented at Figure 5 for melts No. 2 and 4, which are characterized by minimum and maximum strength properties, respectively. The fraction of titanium-based complex carbide in austenite and ferrite in steel No. 4 is higher. This is consistent with TEM results of nanoscale precipitates (Table 3). In addition, (Ti,Nb)C formation in steel No. 4 occurs at a higher temperature. Therefore, for this steel, a greater degree of realization of grain refinement should be expected due to the inhibition of austenite recrystallization caused by the deformation-initiated formation of precipitates, which is observed experimentally. The conditions for the formation of Fe_3C, a complex Fe_7C_3-based carbide, and Mo_2C occur at temperatures below 700 °C. As a result of the diffusion inhibition and the consumption of a significant amount of carbon for the complex titanium-based carbide precipitation, the formation of cementite, molybdenum carbide, and complex iron carbide M_7C_3 is almost completely suppressed.

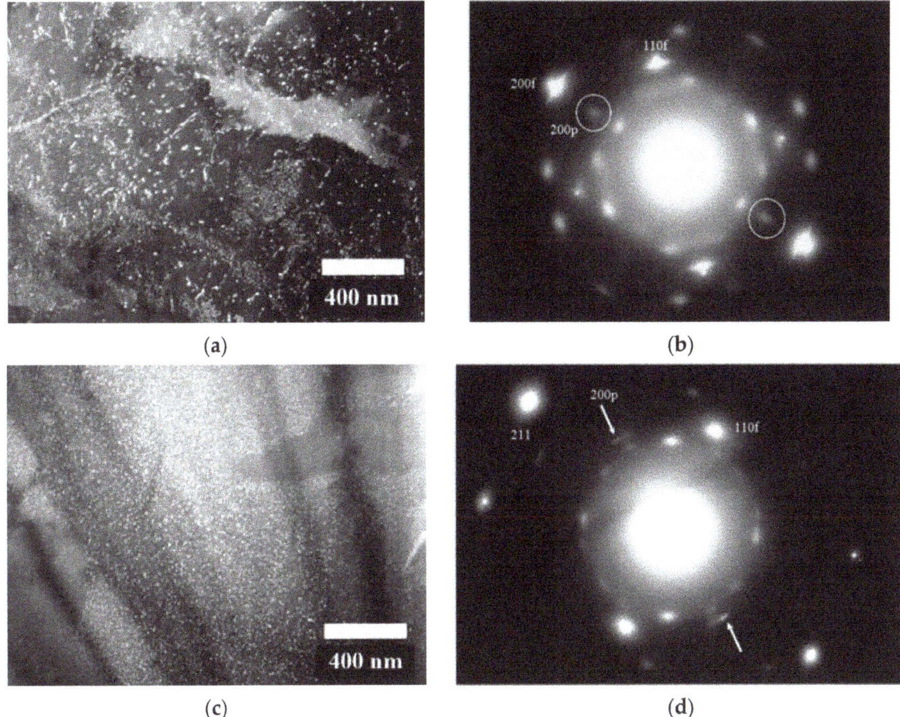

Figure 4. Typical dark-field and microdiffraction images of austenitic precipitates in rolled steel of smelting No. 6 (**a**,**b**) and interphase precipitates in rolled steel of smelting No. 2 (**c**,**d**): (**a**,**c**)—dark-field images, ×30,000, (**b**,**d**)—microdiffraction images.

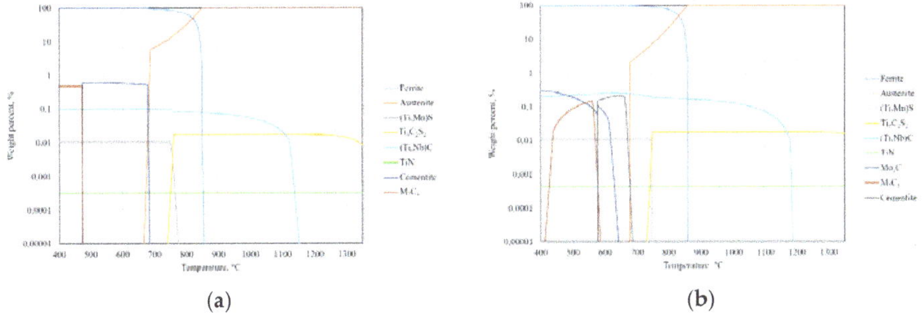

(a) (b)

Figure 5. Temperature dependences of equilibrium fractions of phases in steels No. 2 (**a**) and 4 (**b**). M_7C_3 is a complex Fe_7C_3 -based carbide.

4. Discussion

The results of the present study indicate that the microstructure of complex microalloyed steels of the ferritic class differ from the special features of Ti-Mo steels structure formation. Even at a low cooling rate after the rolling end—10–15 °C/s, block ferrite is formed mainly, although the shape of the blocks is closer to equiaxed, with the exception of smelting No. 6, where the blocks are elongated. Only in rolled products of smelting No. 2, 3 the fraction of PF prevails over share of block ferrite, as in Ti-Mo microalloyed steels [5,20]. This, most likely, is associated with a low content of microalloying elements in the steels under consideration relative to other compositions, including the absence of molybdenum. According to [21], the addition of molybdenum to niobium-containing steels with a manganese content of up to ~1.5 wt.% promotes obtaining of a homogeneous structure of block (acicular) ferrite. It should be noted that microalloying with molybdenum also affects the formation of nanoscale carbide precipitates: their number increases and their size decreases. This is confirmed by a series of studies of steels with different base microalloying components and their combinations [10,22–26]. The favorable effect of molybdenum is explained in [20,27] by the fact that this element can suppress the annihilation of dislocations at high temperature and provide a larger number of carbide nucleation centers. In addition, there is an acceleration of the process of formation of precipitations. According to [20], the incorporation of molybdenum into carbide during the formation of complex precipitates reduces the degree of mismatch between the lattices of the carbide and the matrix due to a decrease in the lattice period, and, consequently, the energy of the carbide/austenite interface. The effect of molybdenum observed in this study is completely consistent with the foregoing.

The amount of austenitic precipitates in steels No. 1, 2, 3 is relatively small compared to Mo microalloyed steels, despite the close carbon content and high Nb concentration—0.11 wt.% in the smelting metal No. 1. In samples of steels containing molybdenum, except steel No. 5, a significant amount of austenitic carbide (carbonitride) precipitates of complex composition containing Nb and V is formed, which is consistent with the data of [23,28]. The number of such precipitates in rolled steel No. 6 is well over due to a significantly higher concentration of vanadium, and in steel of smelting No. 4—titanium. However, the influence of austenitic precipitates on the formation of the structural state and properties of the studied steels is apparently insignificant, since the size of the ferrite grain is in all cases close and, as a rule, is less than 10–15 μm. Only in the case of rolled steels of melts No. 2, 3, 5 in some samples there are grains larger than 15 μm. On the other hand, the amounts of austenitic precipitates detected in steels No. 1, 2, 3 are rather close. Higher strength characteristics and a smaller grain size of rolled steel No. 1 compared to No. 2, most likely, is associated with the inhibition of recrystallization during hot rolling by niobium in solid solution [29], as well as grain refinement during the formation of interphase precipitates [30]. The number of such precipitates in rolled products of steel No. 1 is greater due to higher Nb and C concentrations, low N content and is comparable with this indicator for steels No. 4, 6, 7, 8. This is associated with a significant consumption of microalloying elements, especially in the case of smelting No. 6, for the formation of carbide (carbonitride) precipitates in austenite.

The established special features of changes in the characteristics of the structural state and present phase precipitates have a good correlation with the obtained values of the mechanical properties (Table 2). Rolled steels of smelting No. 1, 6, 7, 8 are characterized by similar strength characteristics. Slightly lower values in the case of smelting No. 2, 3, 5, most likely, are associated with slightly coarser microstructure and fewer interphase precipitates controlling the dispersion hardening. The highest yield and tensile strength were obtained for hot-rolled steel of smelting No. 6 having a complex microalloying system of V, Nb, Ti, and Mo. The presence of high concentrations of V, Mo and an increased Mn content facilitates the transformation of austenite to BF even at a low cooling rate, leading to the formation of an elongated shape of blocks and creates the conditions for the formation of a large number of austenitic and interphase precipitates based on V(C,N) [21,28]. This is in good agreement with the results of [7,23], which showed that a significant contribution to hardening is achieved due to the large number of nanoscale precipitates of vanadium carbide (carbonitride). In this case, a further increase in strength properties is possible with an increase in the cooling rate and carbon content in steel. When a high cooling rate after hot rolling of 50 °C/s was used in [7], the tensile strength of steel with the composition (wt.%): 0.05C-1.7Mn-0.06Nb-0.15Mo-0.22V was 837 MPa. At a high carbon content, hot rolled steel (wt.%): 0.09C-0.29Mn-0.093Ti-0.26Mo-0.14V had a tensile strength of 1000 MPa [23]. It should be noted that the concentrations of microalloying elements present in the steel under consideration are sufficient to bind the present interstitial elements—C, N. As a result, no cementite was detected in the steel microstructure. That means, there is a close analogy with IF steels with a solid solution free of interstitial atoms. The properties of the steels No. 6 and 8 are similar.

In this case, a decrease in the V concentration in the metal of smelting No. 8 is compensated by the increased content of Ti and, to an insignificant degree, Nb and Mn.

The maximum strength properties were obtained for rolled steel of smelting No. 4, characterized by a maximum Ti content and Mo microalloying. As follows from the results of TEM (Table 3) and thermodynamic calculation (Figure 5), high values of yield and tensile strength are the result of grain-boundary and dispersion hardening mechanisms due to the large number of complex nanoscale carbide precipitates (Ti, Nb)C. The niobium content in this steel is small (0.024 wt.%), and the result is in good agreement with the conclusions of the study of Ti-Mo microalloyed steels in [4] on the role of titanium. An increase in the concentration of this element promotes an increase in the number of carbide precipitates. At a higher carbon content (wt.%): 0.08C-1.39Mn-0.21Mo-0.165Ti and a similar regime of thermo-deformation processing, the yield and ultimate strength of 770 and 840 MPa, respectively, were obtained. At the same time, the presence of molybdenum in the composition of the steel stimulates, as noted above, the acceleration of the nucleation of carbide (carbonitride) precipitates, but inhibits their growth. This is well illustrated by the data obtained in the present study. In particular, a simple increase in the Ti content in steel No. 3 led to a less increase in the strength characteristics of rolled products compared to the metal of smelting No. 2 than in the case of smelting No. 4. On the other hand, the amount of carbide (carbonitride) precipitates formed is directly related to the concentration of Ti in the solid solution. Therefore, even if Mo is present in the steel composition, but Ti content is low, it is not possible to form a sufficiently large number of precipitates and to obtain high strength characteristics. This is evidenced by the relatively low yield and tensile strengths of rolled steel of smelting No. 5. It is important that Ti remaining in the solid solution after binding at high temperatures of nitrogen and sulfur can participate in the formation of carbide (carbonitride) precipitates [31], while S and N content in steel of smelting No. 5 is quite high. As a result, only about 0.03 wt.% Ti remains in the solid solution, which is not enough for the effective formation of a sufficiently large number of phase precipitates. A significant consumption of titanium to bind the nitrogen present, in many respects, also leads to a reduced level of strength properties of rolled steel of smelting No. 3. Thus, the presence of increased concentrations of nitrogen and sulfur in the composition of the steels under consideration is an unfavorable factor, since it leads to a decrease in the effectiveness of Ti and Nb. The effect of phosphorus content is less significant, since the formation of complex phosphide precipitates in the general form of FeTiNbP, as well as phosphorus segregation along grain boundaries, for example, as in IF and IF-HS steels at the stage of recrystallization annealing, is of little significance in this case. Therefore, the low phosphorus content obtained in laboratory smelting steels is not a significant factor, since its increase to the level usually found in modern metallurgical practice of 0.006–0.010 wt.% will not change the indicators of the structural state and properties of steel.

It should be noted that Ti and V have the highest propensity for the formation of the most effective interphase carbide precipitates. These alloying metals have a small atomic diameter and, therefore, a greater diffusion mobility [32], which allows the implementation of the process at higher rates of $\gamma \rightarrow \alpha$ transformation of steel. Moreover, Ti has a smaller atomic diameter, higher thermodynamic stability, and, hence, the higher driving force for the formation of carbide precipitates in steel. This, in particular, is indicated by the results of calculating the stability conditions for the carbide, nitride, and carbonitride phases of the considered steel elements [30,33]. The formation of this type of titanium precipitates occurs at incomparably higher temperatures than vanadium under other equivalent conditions. The noted circumstances correlate well with the results obtained. From the data of Tables 1 and 2, it can be seen that rolled products produced under equivalent conditions (steels No. 6 and 8) have higher strength characteristics with a lower titanium content compared with vanadium. The participation of niobium in the formation of interphase precipitates is greatly limited due to the significantly larger size and lower diffusion mobility of atoms despite the high thermodynamic stability of carbonitride phases comparable with titanium [30,32]. Values of the relative elongation of the studied rolled steels are quite close. However, steels, which containing titanium as the main microalloying element (having the highest concentration), in general, have slightly higher values than in the case of vanadium. Taking into account economic and cost indicators, the results obtained testify to the advantages of titanium-based microalloying systems in comparison with vanadium-based.

5. Conclusions

Thus, a study of the microstructure, phase precipitates, and mechanical properties of separately or jointly V, Nb, Ti, and Mo microalloyed low-carbon hot-rolled steels of the ferritic class made it possible to establish the following special features. When they are cooled after rolling, even at a low cooling rate of 10–15 °C/s, unlike Ti, Mo microalloyed steels, block ferrite is formed mainly. The shape of the blocks is closer to equiaxial, except for steel with a high content of microalloying components, where the blocks have an elongated shape. On the contrary, in low microalloyed steels, the fraction of block ferrite is only ~20–35%, and PF predominates, as in Ti-Mo steels. In the metal of all melts, the presence of cementite precipitates up to several microns in size was detected, except for steel with high Ti and V content.

The amount of austenitic carbide (carbonitride) precipitates in Nb, Ti microalloyed steels is much smaller than in the case of complex V, Nb, Ti, and Mo microalloying, except for steel with a low V, Nb, and Ti content. Their effect on the structural state and properties of steel is negligible. The presence of more efficient nanoscale interphase/mixed type precipitates was detected in all steels, except steel, characterized by low V, Nb, and Ti concentrations. In different areas of the metal, the ratio of volume fractions of interphase and austenitic precipitates can vary significantly. In those grains (blocks) where both types of precipitations are located, there are areas occupied exclusively by precipitations of one type or another. The noted circumstance shows that the formation of austenitic precipitates significantly reduces the potential of microalloying elements and prevents the formation of interphase precipitates. The data obtained by TEM methods on the number of various types of phase precipitates are confirmed by the results of a thermodynamic calculation of the temperature dependences of the equilibrium phase composition of steels.

The established special features of changes in the structural state and present phase precipitates correlate well with the obtained values of the mechanical properties. As in Ti-Mo microalloyed steels, the strength characteristics of rolled products enhance upon obtaining the microstructure of block ferrite and an increase in the number of nanoscale interphase and mixed type precipitates, the amount of which grows with increasing V, Nb, and Ti concentrations and the presence of Mo in the metal composition. In addition, the highest yield and tensile strengths were obtained for hot-rolled Nb, Ti, and Mo microalloyed steel with a significant titanium content of 0.17 wt.%. It is shown that rolled products obtained under equivalent conditions have higher strength characteristics at a lower titanium content compared to vanadium. Values of the relative elongation of the studied rolled steels are quite close. However, for steels with the titanium-based microalloying system, slightly higher values are obtained than in the case of vanadium-based. Taking into account economic indicators, the obtained results testify to the advantages of using titanium-based microalloying systems for the investigated type of steels in comparison with vanadium-based.

Author Contributions: Conceptualization, A.Z. and A.K.; methodology, A.Z. and A.K.; software, A.K.; validation, A.Z. and S.D.; formal analysis, S.D.; investigation, A.K., N.A. and D.D.; writing—original draft preparation, N.A.; writing—review and editing, A.Z.; supervision, A.Z.; project administration, A.Z. All authors have read and agreed to the published version of the manuscript.

Funding: This research was supported by the grant of the Russian Science Foundation (Project No. 18-19-00639) and was performed at Bardin Central Research Institute of Ferrous Metallurgy.

Conflicts of Interest: The authors declare no conflict of interest.

References

1. Lesch, C.; Kwiaton, N.; Klose, F.B. Advanced high strength steels (AHSS) for automotive applications—Tailored properties by smart microstructural adjustments. *Steel Res. Int.* **2017**, *88*. [CrossRef]
2. Fonstein, N. *Advanced High Strength Sheet Steels*; Springer Science and Business Media LLC: Cham, Switzerland, 2015; pp. 3–14.

3. Funakawa, Y.; Shiozaki, T.; Tomita, K.; Yamamoto, T.; Maeda, E. Development of High Strength Hot-Rolled Sheet Steel Consisting of Ferrite and Nanometer-Sized Carbides. *ISIJ Int.* **2004**, *44*, 1945–1951. [CrossRef]
4. Koldaev, A.V.; Zaitsev, A.I.; Krasnyanskaya, I.A.; D'yakonov, D.L. Study of the effect of composition and thermal deformation treatment on properties of ferritic steels microalloyed with titanium and niobium. Part 2. Phase precipitate characteristics. *Metallurgist* **2019**, *63*, 604–616.
5. Koldaev, A.V.; Zaitsev, A.I.; Krasnyanskaya, I.A.; D'Yakonov, D.L. Study of the Effect of Composition and Thermal Deformation Treatment on Properties of Ferritic Steels Microalloyed with Titanium and Niobium. Part 1. Microstructure Characteristics. *Metallurgist* **2019**, *63*, 487–495. [CrossRef]
6. Deng, X.; Fu, T.; Wang, Z.; Liu, G.; Wang, G.; Misra, R.D.K. Extending the boundaries of mechanical properties of Ti-Nb low-Carbon steel via combination of ultrafast cooling and deformation during austenite-to-Ferrite transformation. *Met. Mater. Int.* **2017**, *23*, 175–183. [CrossRef]
7. Rijkenberg, A.; Blowey, A.; Bellina, P.; Wooffindin, C. Advanced high stretch-flange formability steels for chassis & suspension applications. In Proceedings of the SCT2014 (4th International Conference on Steels in Cars and Trucks), Braunschweig, Germany, 15–19 June 2014; pp. 426–433.
8. Funakawa, Y.; Seto, K. Coarsening Behavior of Nanometer-Sized Carbides in Hot-Rolled High Strength Sheet Steel. *Mater. Sci. Forum.* **2007**, *539*, 4813–4818. [CrossRef]
9. Jia, Z.; Misra, R.; Omalley, R.; Jansto, S. Fine-Scale precipitation and mechanical properties of thin slab processed titanium–niobium bearing high strength steels. *Mater. Sci. Eng. A* **2011**, *528*, 7077–7083. [CrossRef]
10. Bu, F.; Wang, X.; Yang, S.; Shang, C.; Misra, D. Contribution of interphase precipitation on yield strength in thermomechanically simulated Ti–Nb and Ti–Nb–Mo microalloyed steels. *Mater. Sci. Eng. A* **2015**, *620*, 22–29. [CrossRef]
11. Chen, C.; Yen, H.-W.; Kao, F.; Li, W.; Huang, C.; Yang, J.-R.; Wang, S. Precipitation hardening of high-Strength low-Alloy steels by nanometer-Sized carbides. *Mater. Sci. Eng. A* **2009**, *499*, 162–166. [CrossRef]
12. Seto, K.; Funakawa, Y.; Kaneko, S. Hot Rolling High Strength Steels for Suspension and Chassis Parts "NANOHITEN" and "BTH Steels". *JFE Tech. Rep.* **2007**, *10*, 19–25.
13. Mao, X.; Huo, X.; Sun, X.; Chai, Y. Strengthening mechanisms of a new 700MPa hot rolled Ti-Microalloyed steel produced by compact strip production. *J. Mater. Process. Technol.* **2010**, *210*, 1660–1666. [CrossRef]
14. Frisk, K.; Borggren, U. Precipitation in Microalloyed Steel by Model Alloy Experiments and Thermodynamic Calculations. *Met. Mater. Trans. A* **2016**, *47*, 4806–4817. [CrossRef]
15. Kamikawa, N.; Sato, K.; Miyamoto, G.; Murayama, M.; Sekido, N.; Tsuzaki, K.; Furuhara, T. Stress–Strain behavior of ferrite and bainite with nano-Precipitation in low carbon steels. *Acta Mater.* **2015**, *83*, 383–396. [CrossRef]
16. Shaposhnikov, N.G.; Koldaev, A.V.; Zaitsev, A.I.; Rodionova, I.G.; D'yakonov, D.L.; Arutyunyan, N.A. Regularities of titanium carbide precipitation in low carbon Ti–Mo-Microalloyed high strength steels. *Metallurgist* **2016**, *60*, 810–816. [CrossRef]
17. Zaitsev, A.I.; Rodionova, I.G.A.V.; Arutyunyan, N.A.; Dunaev, S.F. Investigation of regularities of phase precipitation formation, structural state, and properties of microalloyed low-carbon steels of the ferritic class. *Metallurgist* **2020**, *64*. in press.
18. Khalid, F.A.; Edmonds, D.V. Interphase precipitation in microalloyed engineering steels and model alloy. *Mater. Sci. Eng.* **1993**, *9*, 384–396. [CrossRef]
19. Shaposhnikov, N.G.; Rodionova, I.A.; Pavlov, A.A. Thermodynamic Development of Austenite-Martensite Class Corrosion-Resistant Steels Intended for a Bimetal Cladding Layer. *Metallurgist* **2016**, *59*, 1195–1200. [CrossRef]
20. Chen, C.Y.; Chen, C.C.; Yang, J.-R. Microstructure characterization of nanometer carbides heterogeneous precipitation in Ti–Nb and Ti–Nb–Mo steel. *Mater. Charact.* **2014**, *88*, 69–79. [CrossRef]
21. Xinjun, S. The roles and applications of molybdenum element in low alloy steels. In Proceedings of the International Seminar on Applications Mo in Steels, Beijing, China, 27–28 June 2010; pp. 60–74.
22. Wang, Z.Q.; Zhang, H.; Guo, C.; Liu, W.; Yang, Z.; Sun, X.; Zhang, Z.; Jiang, F. Effect of molybdenum addition on the precipitation of carbides in the austenite matrix of titanium micro-alloyed steels. *J. Mater. Sci.* **2016**, *51*, 4996–5007. [CrossRef]
23. Zhang, K.; Li, Z.-D.; Sun, X.-J.; Yong, Q.-L.; Yang, J.-W.; Li, Y.-M.; Zhao, P.-L. Development of Ti–V–Mo Complex Microalloyed Hot-Rolled 900-MPa-Grade High-Strength Steel. *Acta Met. Sin. Engl. Lett.* **2015**, *28*, 641–648. [CrossRef]

24. Kamikawa, N.; Abe, Y.; Miyamoto, G.; Funakawa, Y.; Furuhara, T. Tensile Behavior of Ti,Mo-Added Low Carbon Steels with Interphase Precipitation. *ISIJ Int.* **2014**, *54*, 212–221. [CrossRef]
25. Zhang, Z.; Yong, Q.; Sun, X.; Li, Z.; Wang, Z.; Zhou, S.; Wang, G. Effect of Mo addition on the precipitation behavior of carbide in Nb-bearing HSLA steel. In Proceedings of the Conference HSLA Steels 2015, Microalloying 2015 & Offshore Engineering Steels 2015, Hangzhou, China, 11–13 November 2015; pp. 203–210.
26. Jang, J.H.; Heo, Y.-U.; Lee, C.-H.; Bhadeshia, H.; Suh, D.-W. Interphase precipitation in Ti–Nb and Ti–Nb–Mo bearing steel. *Mater. Sci. Technol.* **2013**, *29*, 309–313. [CrossRef]
27. Lee, W.; Hong, S.; Park, C.; Kim, K.; Park, S. Influence of Mo on precipitation hardening in hot rolled HSLA steels containing Nb. *Scr. Mater.* **2000**, *43*, 319–324. [CrossRef]
28. Zaitsev, A.I.; Koldaev, A.V.; Krasnyanskaya, I.A.; Dunaev, S. Study of Features of Phase Precipitate Formation, Structural State, and Properties of Nb, V-Microalloyed Low Carbon Ferritic Steels. *Metallurgist* **2020**, *63*, 1033–1042. [CrossRef]
29. Hutchinson, C.R.; Zurob, H.; Sinclair, C.; Brechet, Y. The comparative effectiveness of Nb solute and NbC precipitates at impeding grain-boundary motion in Nb steels. *Scr. Mater.* **2008**, *59*, 635–637. [CrossRef]
30. Lagneborg, R.; Siwecki, T.; Zajac, S.; Hutchinson, B. The role of vanadium in microalloyed steels. *Scand. J. Metall.* **1999**, *28*, 1–86.
31. Zaitsev, A.I.; Rodionova, I.G.; Koldaev, A.V.; Arutyunyan, N.A. Effect of composition and processing parameters on the microstructure and mechanical properties of cold-Rolled and galvanized roll products from IF steels. *Metallurgist* **2020**, *64*, 136–144. [CrossRef]
32. Oono, N.; Nitta, H.; Iijima, Y. Diffusion of niobium in α-Iron. *Mater. Trans. JIM* **2003**, *44*, 2078–2083. [CrossRef]
33. Zaitsev, A.I.; Koldaev, A.V.; Gladchenkova, Y.S.; Shaposhnikov, N.G.; Dunaev, S.F. Structural State Evolution and Rolled Product Properties in Relation to Treatment Regime of Model Steels for Hot Stamping. 1. Hot-Rolled Product. *Metallurgist* **2016**, *60*, 274–280. [CrossRef]

Article

Spark Plasma Sintering of Cobalt Powders in Conjunction with High Energy Mechanical Treatment and Nanomodification

Van Minh Nguyen [1], Rita Khanna [2,*], Yuri Konyukhov [3], Tien Hiep Nguyen [3], Igor Burmistrov [4], Vera Levina [3], Ilya Golov [3] and Gopalu Karunakaran [5]

1 Institute of Research and Development, Duy Tan University, Danang 550000, Vietnam; chinhnhan88@gmail.com
2 School of Materials Science and Engineering (Ret.), University of New South Wales (UNSW), Kensington, Sydney 2052, Australia
3 Department of Functional Nanosystems and High-Temperature Materials, National University of Science and Technology "MISiS", Moscow 119049, Russia; martensit@mail.ru (Y.K.); htnru7@yandex.ru (T.H.N.); vvlevina@gmail.com (V.L.); gorlym@live.ru (I.G.)
4 Engineering Centre, Plekhanov Russian University of Economics, Moscow 117997, Russia; glas100@yandex.ru
5 Biosensor Research Institute, Department of Fine Chemistry, Seoul National University of Science and Technology (Seoul Tech.), Seoul 232, Korea; karunakarang5@gmail.com
* Correspondence: rita.khanna66@gmail.com; Tel.: +82-61-04217-97500

Received: 24 April 2020; Accepted: 21 May 2020; Published: 23 May 2020

Abstract: Spark plasma sintering (SPS) investigations were carried out on three sets of Co specimens: untreated, high energy mechanically (HEMT) pre-treated, and nanomodified powders. The microstructure, density, and mechanical properties of sintered pellets were investigated as a function of various pre-treatments and sintering temperatures (700–1000 °C). Fine-grained sinters were obtained for pre-treated Co powders; nano-additives tended to inhibit grain growth by reinforcing particles at grain boundaries and limiting grain-boundary movement. High degree of compaction was also achieved with relative densities of sintered Co pellets ranging between 95.2% and 99.6%. A direct co-relation was observed between the mechanical properties and densities of sintered Co pellets. For a comparable sinter quality, sintering temperatures for pre-treated powders were lower by 100 °C as compared to untreated powders. Highest values of bending strength (1997 MPa), microhardness (305 MPa), and relative density (99.6%) were observed for nanomodified HEMT and SPS processed Co pellets, sintered at 700 °C.

Keywords: spark plasma sintering; nanomodification; mechanical processing; densification; mechanical properties

1. Introduction

Spark plasma sintering is finding increasing application as a key technology for hard or very hard to sinter high performance materials [1,2]. The SPS technique has made significant contributions to the development of advanced materials such as refractory materials, functional nano ceramics, non-equilibrium materials, biomaterials, etc. [3]. Based on low voltage, pressure assisted, fast heating with DC pulsed current, the SPS technique has high sintering speed, reproducibility, and reliability with a great potential for achieving fast densification results with minimal grain growth in a short sintering time [4–7]. Depending on the experimental configuration, sample geometry, thermal and electrical characteristics, heating rates as high as 1000 °C/min and processing times of few minutes can be achieved [4]. A few representative examples of SPS application include: the strengthening of alumina matrix with nanocomposites [8], large sized bulk metallic glasses [9,10], consolidation of metallic powders [11–13], formation of porous materials [14], biomedical applications [15], formation of super alloys [16], carbides [17], orthopedic alloys [18], etc.

During SPS operations, the heating power is distributed homogeneously over the volume of the powder compact on a macroscopic scale. However, microscopically the heating power is dissipated preferentially at the contact points between powder particles [4]. Pulsed current causes surface activation of powders and enhanced diffusion in contact areas. According to the plasma/micro-spark theory, a spark discharge occurring between powder particles creates a local high temperature state leading to vaporization, local surface melting, and the formation of necks/constricted shapes in the contact region [19]. As only the surface temperatures rise sharply through self-heating, the grain growth, coarsening, and particle decomposition of the original particle are inhibited during the SPS process. [20] For example, nanosized particles can be sintered using SPS without considerable grain growth [21]. Such consolidation behavior can also result in completely novel microstructures and materials, thereby opening new areas of research and applications. Due to plastic transformations, high relative densities for powder sinters can be achieved in a short time.

Deng et al. (2019) reported on minor non-thermal effects of electric current such as electromigration-enhanced densification during the SPS process [22]. Lee et al. [2] reported on the densification mechanisms for SPS consolidation of tungsten powders; the sintering was carried in the temperature range 1600–1800 °C (melting point of W: 3422 °C) and pressures of 60 and 120 MPa. The relative density of the compacted pellets was found to range between 81–95%; there was some evidence for superplastic-like behavior based on electron back-scattering results. Knaislová et al. [23] carried out high pressure SPS sintering of Ti-Al-Si intermetallic alloys at pressures of 6 GPa and temperatures ranging from 1045 to 1324 °C to produce low-porosity consolidated specimens of Ti_5Si_3 silicides in an TiAl matrix. Marek et al. (2016) reported on the synthesis of ultrafine and microcrystalline cobalt using high-energy ball milling followed by the SPS process; relative densities of 88.8% to 97.8% were achieved [24]. Kundu et al. (2019) used high energy ball milling on Fe-9Cr alloys followed by SPS consolidation for a range of dwell times (7–45 min) and temperatures (850–1050 °C) to achieve a maximum relative density of 98% [25].

The aim of this study is to enhance the efficacy of the SPS process with appropriate pre-treatments, and to investigate the densification, mechanical and microstructural properties of sintered cobalt pellets. Due to high corrosion resistance, excellent mechanical properties, biocompatibility, cobalt-based alloys are used in the manufacture of combustion chambers in gas turbines [26], strengthened superalloys [27], biomedical and orthopedic applications, surgical implants, etc. [28,29]. Using high-energy mechanical treatment and nanomodification in conjunction with the SPS process, in-depth investigations were carried out on three sets of specimens over a range of temperatures. The influence of various operating conditions on density, microstructure, and mechanical properties of sintered Co pellets was investigated towards process optimization, and to produce high-quality sintered materials at lower sintering temperatures.

The article is organized as follows. Details on the synthesis of cobalt nanopowders, operational details of high energy mechanically and SPS treatments are presented in Section 2 (Materials and Methods), along with a detailed experimental plan. Results on density measurements, microstructure, bending strength, microhardness, and carbon diffusion on the reacted surface are presented in Section 3 (Results and Discussion), followed by critical analysis and discussion. Key findings from the study and conclusions are summarized in Section 4 (Conclusions).

2. Materials and Methods

There were three key experimental procedures in this study: synthesis of Co nanopowders, high energy mechanical treatment, and spark plasma sintering. Operational features of these procedures and experimental plan are detailed below.

2.1. *Synthesis of Co Nanopowders*

Co micropowder (Brand: PK-lu-electrolytic, NPO Rusredmet, St. Petersburg, Russia) was used as the starting material for this investigation. Basic characteristics of Co powder were determined as: purity ≥ 99.35%; Brunauer-Emmet-Teller (BET) surface area: 0.2 m^2/g; average particle size: 3.4 μm. Chemical precipitation was used to synthesize nanoparticles using the precipitation of Co $(OH)_2$ from 10% Co $(NO_3)_2$; 10% NaOH solution was added to catalyze precipitation. This reaction was followed by thermal reduction at 265 °C for 3 h under H_2 flow, and the resulting nanopowders were passivated at liquid nitrogen temperatures. The presence of a small amount of oxygen in the commercial grade technical N_2 (<0.4% oxygen) can lead to the formation of a thin (up to 5 nm) oxide surface layer on nanoparticles. Therefore, these nanopowders could contain up to 3 wt. % of oxygen. Synthesized Co nanoparticles were characterized using advanced analytical techniques: Scanning electron microscopy (Tescan Vega 3, TESCAN, Brno, Czech Republic); transmission electron microscopy (JEM-2010, JEOL, Tokyo, Japan); X-ray diffraction (Difrey 401, Scientific Instruments, Russia), BET surface area measurement (NOVA 1200, Quantachrome Instruments, Boynton Beach, Florida, USA). Detailed characterization results are presented next.

Figure 1A shows X-ray diffraction results (Cu Kα; 40 kV; 40 mA) for synthesized Co nanopowders. Most of the diffraction peaks (shown in 'black') corresponded to the hcp phase of Co. The observed peak height of the (101) peak (100%; 47.39°) was much smaller than the (002) peak (28.6%; 44.26°). The (101) peak was also found to be relatively broad, whereas (002) peak was quite sharp and narrow. These results indicate the presence of texture in these nanopowders. A small amount of the fcc phase was also detected with two weak peaks (shown in 'red') at 45.922° ((111); 100%) and 53.456° ((002); 41.9%). Figure 1B shows a selected area electron diffraction patten (SAED) on cobalt nanoparticles during transmission electron microscopy (TEM) investigations. The presence of sharp diffraction spots indicates the crystalline nature of these particles. This result is in good agreement with X-ray diffraction results (Figure 1A).

A.

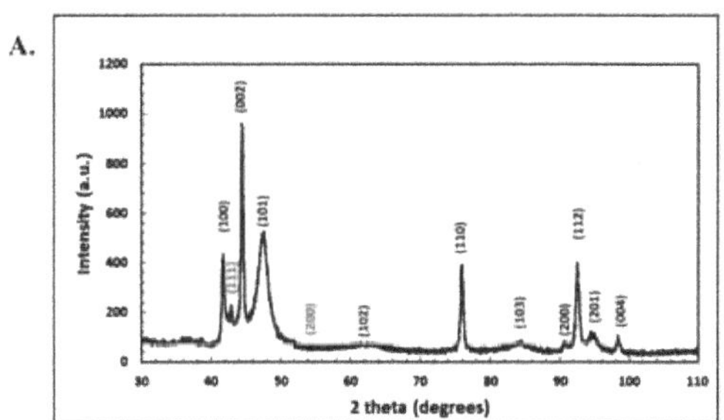

B.

Figure 1. (**A**) X-ray diffraction (Cu Kα), and (**B**) selected area electron diffraction (SAED) results on synthesized Co nanopowders.

Figure 2a–d shows electron microscopy images (both SEM and TEM) of synthesized Co nanoparticles in a range of magnifications. While most of the particles were well rounded, their shapes were generally irregular with no well-defined form or shape. There was no specific tendency towards spherical morphologies. These nanoparticles tended to reduce their surface area by agglomerating into clumps larger than 150–200 nm (Figure 2a–d). From BET surface area measurements, the specific surface area of synthesized Co nanoparticles was determined to be 10.1 m^2/g. The average particle size was found to be 67 nm. The presence of agglomerates imposes certain requirements for their introduction into polydisperse media. Additional energy needs to be supplied to the system to break up these aggregates and to achieve a high degree of homogenization of the mixture during the nanoparticle injection process.

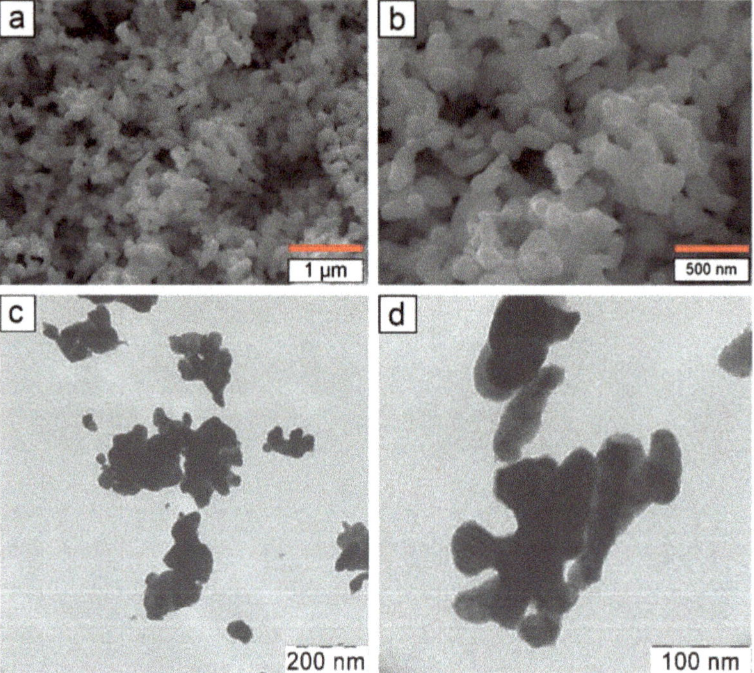

Figure 2. (**a**,**b**) SEM and (**c**,**d**) TEM micrographs of synthesized Co nanopowders.

2.2. High Energy Mechanical Treatment

A magnetic mill (UAP-3, 21st Century Advanced Technology, Russia) was used to carry out HEMT on Co powders under argon atmosphere. Key components of the equipment include a magnetic inductor to generate a high-power vortex magnetic field, water cooling circuit, electric resistance tube furnace, and a steel reaction chamber (45 mm dia., 350 mm length). Further details have been given elsewhere [30]. The simultaneous influence of high-power vortex magnetic field (0.16 T) and mechanical processing of sample powders with ferromagnetic stainless steel needles in the reaction chamber creates intense vortex rotation with speeds reaching 3200 rpm. Pressures within the chamber can build up to several thousand MPa during operation. In this study, hard-wearing steel needles of length 15–20 mm, 0.6–1.2 mm dia. were used as the ferromagnetic bodies to form the vortex layer. The relative proportion of ferromagnetic needles to Co powder in the reaction chamber was chosen to be 4:1 by weight. Co nanoparticles (0.5 wt. %) were first mixed with Co micro-powders in air.

The HEMT treatment was carried out on the mixture in an argon atmosphere. Steel needles were separated from the processed material by a sieve. A processing time of 3 min was used for all powders.

During the HEMT operation, sample powders get mixed extensively in the reaction chamber and may undergo deformations along with changes in structure and reactivity [30,31]. Figure 3 shows the morphology of Co particles before and after HEMT. Co particles were almost spherical initially (Figure 3a). One minute of HEMT processing shows surface deformation and elongation of some particles; not all particles were equally affected (Figure 3b). However, 2 min of HEMT processing showed a significant flattening, increases in particle sizes, and transformation from spherical to disc shapes (Figure 3c). Such changes in morphology can play a significant role in increasing contact regions between Co particles during sintering.

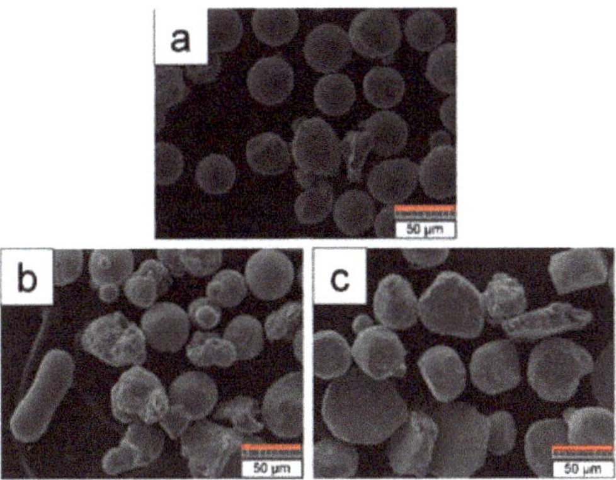

Figure 3. SEM images of Co powders after HEMT operation. (**a**) t = 0 min, (**b**) t = 1 min, (**c**) t = 2 min.

2.3. Spark Plasma Sintering

Commercial SPS equipment LABOX-650 (Sinter Land Inc., Kyoto, Japan) was used in this investigation. Key operating parameters were temperature: ≤2400 °C; pressing force: 0.5–60 kN; heating rates: ≤500 °C/min; operating current: ≤5000 A; cylinder stroke 150 mm; sintering under vacuum or inert atmosphere. Approximately 27 g of cobalt powder was loaded into each matrix with an internal diameter of 3 cm. The inside of matrix was covered by graphite sheets to facilitate the removal of the sintered product. Sintering of powders was carried out under pressure and high temperature spark plasma, generated as an electric discharge between the particles under the influence of a powerful pulsed current. Operating parameters during sintering were pressure 50 MPa, heating rate 50 °C/min, holding time 5 min, temperatures ranging from 700 to 1000 °C. A pulsed direct current (500–2500 A) was passed through the sample and the mold; the temperature was measured with help of a thermocouple and a pyrometer. The voltage and current were changed automatically to provide the preset heating and holding modes. Sintered samples were cooled to room temperature. The average height of sintered cobalt pellets was found to be 4.2 ± 0.1 mm.

2.4. Experimental Plan

An outline of the experimental plan is shown in Figure 4. Spark plasma sintering was carried out on three sets of specimens: (a) un-treated Co micro-powders, (b) HEMT processed Co powders, and (c) nanomodified (0.5 wt. % Co nanopowders additive) Co micro-powders subjected to HEMT processing. The concentration of nanopowders in the blend was determined by technological and economic considerations [32–35]. Detailed structure–property measurements including density, microstructure, microhardness, and bending strengths were determined for Co pellets sintered for a range of temperatures.

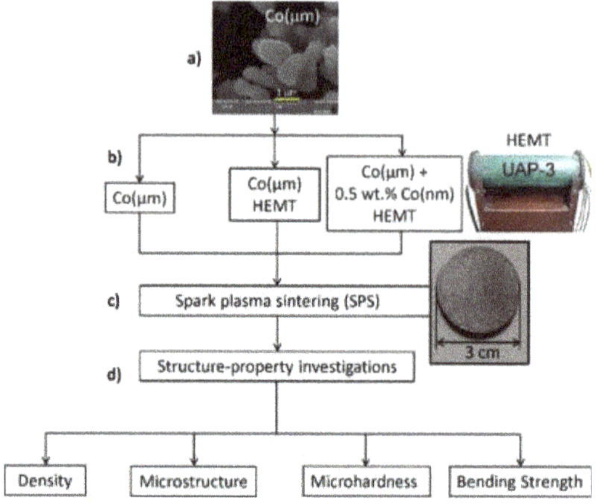

Figure 4. An overview of the experimental plan.

3. Results and Discussion

The SPS consolidation was carried out on three sets of Co powders over a range of temperatures. Detailed results on sintered pellets are presented next. These are followed by critical analysis and discussion.

3.1. Morphology and Microstructure

High resolution micrographs depicting the morphologies of Co powders following HEMT and nanomodification pre-treatments are shown in Figure 5. Figure 5a (set I) shows that untreated, as received Co particles were well rounded with sizes generally around 3 μm, which is consistent with BET results on starting Co powders. The HEMT processing (Figure 5b, set II) led to the crushing, milling, surface roughening, and size reduction of Co microparticles producing irregular shaped particles in a range of sizes. Figure 5c (set III) shows the effect of HEMT processing on nanomodified Co powders. Nanoparticles were seen evenly distributed throughout the powder mixture and showed a tendency to attach with bigger microparticles. The HEMT pre-treatment was seen to cause a high degree of deformation, mixing, homogenization, and activation of Co powders prior to SPS consolidation.

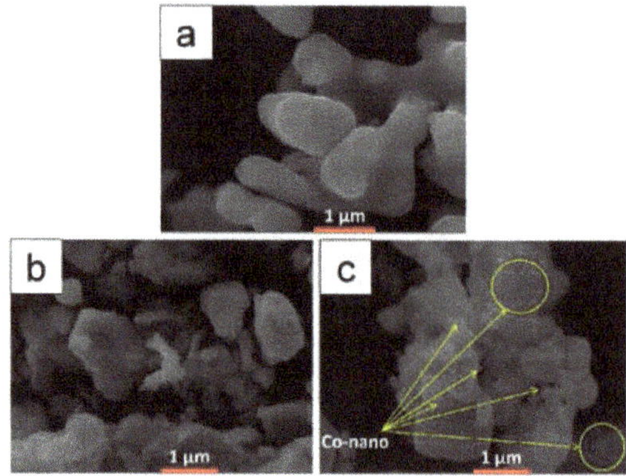

Figure 5. The SEM micrographs of three sets of Co micropowders: (**a**) untreated Co powder, (**b**) HEMT treated Co powder, (**c**) nanomodification, and HEMT treatments on Co powders.

Figure 6 shows optical micrographs (Axio Observer D1m; Carl Zeiss, Oberkochen, Germany) on the microstructure of Co pellets sintered at 700 °C. Samples were polished to a mirror finish and were etched with 30 wt.% nitric acid for 5 s. The largest grain size was observed in Figure 6a (set I) for Co powders without any pre-treatment. Figure 6b,c (sets II and III) show that the grain-size had decreased significantly for HEMT processed and nanomodified and HEMT treated Co powders. The largest number of grains were observed for set III.

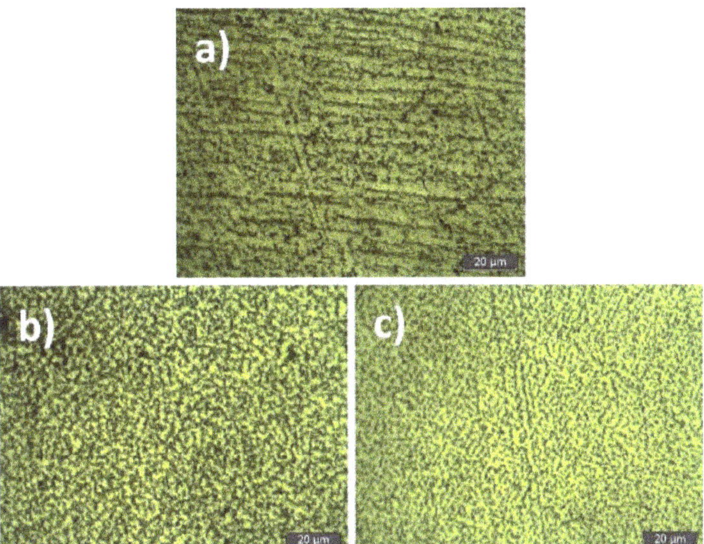

Figure 6. Microstructure of sintered Co pellets (700 °C) for sets (**a**) I, (**b**) II, (**c**) III of Co powders as determined through optical microscopy.

3.2. Densification

The densities of sintered samples were determined using gas pycnometry technique (Ultra pycnometer 1000, Quantachrome Instruments, Boynton Beach, Florida, USA). These measurements were made with an accuracy of ±0.03%. Detailed results are shown in Figure 7. Figure 7A shows the impact of sintering temperatures on the three sets of specimens. Densities of sintered Co pellets were found to range between 8473 to 8860 Kg/m^3 (theoretical density: 8890 Kg/m^3). Corresponding relative densities have been plotted in Figure 7B. These were found to range between 95.2% to 99.6%, indicating extensive densification under certain operating conditions.

For the set I representing SPS consolidation on as received Co micro-powders, the density of sintered pellets increased from 8585 Kg/m^3 (700 °C) to 8840 Kg/m^3 (800 °C) and later decreased with increasing temperatures reaching magnitudes of 8528 Kg/m^3 (900 °C) and 8604 Kg/m^3 (1000 °C). Maximum relative density of 99.3% was achieved at 800 °C. For the set II representing SPS results on Co micro-powders with HEMT pre-treatment, highest pellet density (8847 Kg/m^3) was achieved at 700 °C. Higher processing temperatures showed lower densities: 8660 Kg/m^3 (800 °C), 8473 Kg/m^3 (900 °C), 8615 Kg/m^3 (1000 °C). Results for the set III, representing SPS results on nanomodified Co micro-powders with HEMT pre-treatment, followed a trend similar to the set II. However, overall densities were slightly higher in set III as compared to set II. Highest pellet density (8860 Kg/m^3) was achieved at 700 °C. Higher processing temperatures generally showed lower densities: 8740 Kg/m^3 (800 °C), 8556 Kg/m^3 (900 °C), 8689 Kg/m^3 (1000 °C).

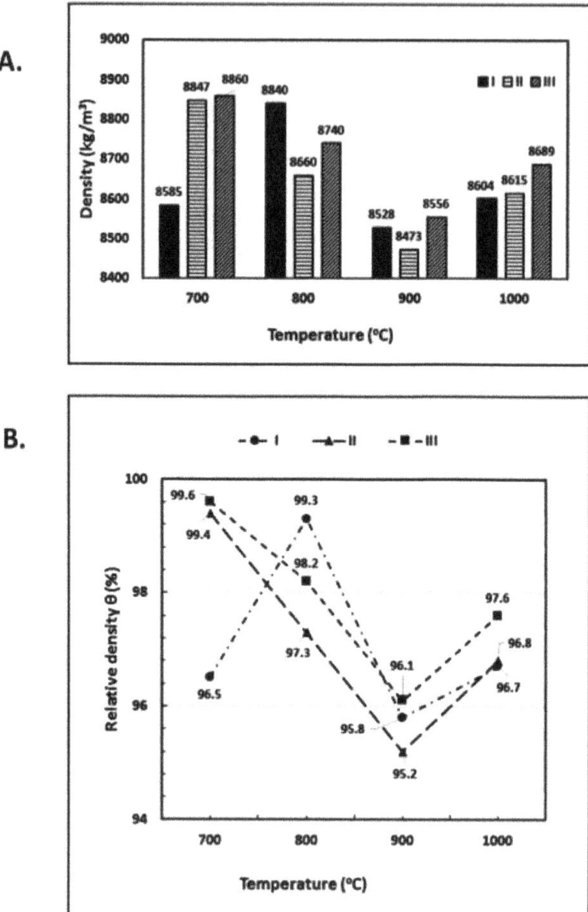

Figure 7. Influence of sintering temperature on (**A**) absolute densities and (**B**) relative densities of three sets of Co powders.

3.3. Mechanical Properties

The mechanical properties of sintered samples such as microhardness and bending strength were determined using the microhardness tester (Tukon TM 1102, Wilson hardness, Norwood, MA, USA) and universal testing machine (Instron 5966, Instron, Norwood, MA, USA), respectively. The impact of pre-treatments and sintering temperatures on the bending strength of sintered pellets is shown in Figure 8a. At 700 °C, the bending strength was lowest for untreated Co powders (set I: 1639 MPa). With HEMT pre-treatment and nanomodification, increases to 1878 MPa (set II) and 1997 MPa (set III) were recorded. This trend was reversed at 800 °C, albeit with lower bending strengths, with highest value recorded for set I (1781 MPa); sets II and III recorded 1679 and 1728 MPa, respectively. All three sets showed similar bending strengths at 900 °C. Bending strengths showed a small increase at 1000 °C with trends between various sets similar to those observed at 700 °C. Bending strengths of sintered Co pellets were strongly affected by the initial pre-treatments and temperatures of SPS consolidation. Maximum bending strength of 1997 MPa was recorded for set III pellets sintered at 700 °C.

The impact of pre-treatments and sintering temperatures on the microhardness of sintered Co pellets is shown in Figure 8b. The microhardness of sintered Co pellets was determined using Vickers scale with a load of 0.3 kgf. There appears to be a direct correspondence between the bending strength and microhardness results. At 700 °C, the microhardness was lowest for untreated Co powders (set I: 271 MPa). With HEMT pre-treatment and nanomodification, increases to 298 MPa (set II) and 305 MPa (set III) were recorded. All three sets showed similar microhardness values at 800 °C; while set I showed a marginal increase, small reductions were observed for both sets II and III. Further reductions were observed at 900 °C, followed by a small increase at 1000 °C. Following a trend similar to bending strength, microhardness of sintered Co pellets showed a strong dependence on initial pre-treatments and temperatures of SPS consolidation. Highest values of microhardness were observed for nanomodified HEMT and SPS processed samples (set III). Maximum hardness of 305 MPa was recorded for set III at 700 °C.

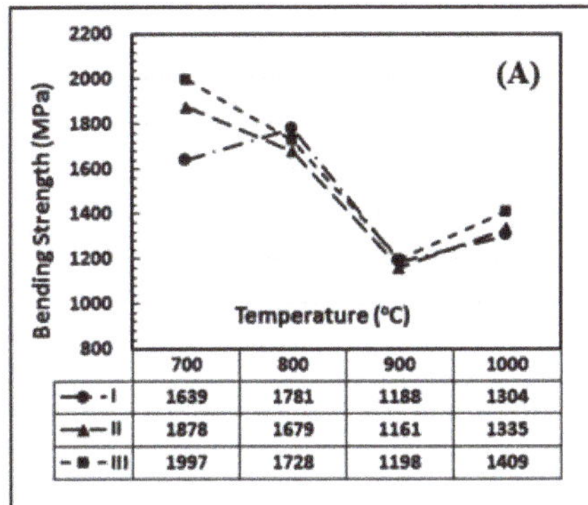

	700	800	900	1000
I	1639	1781	1188	1304
II	1878	1679	1161	1335
III	1997	1728	1198	1409

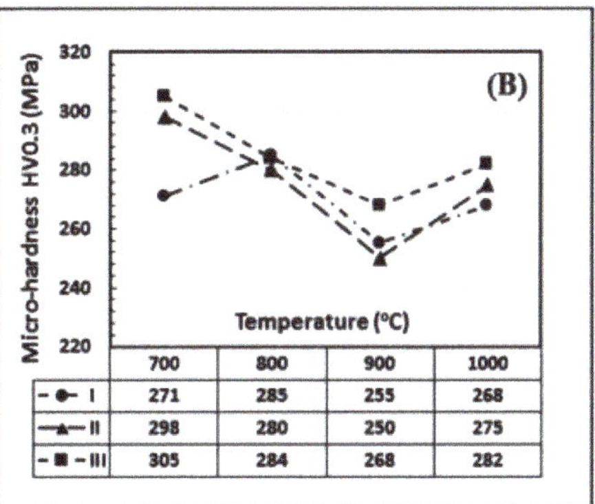

	700	800	900	1000
I	271	285	255	268
II	298	280	250	275
III	305	284	268	282

Figure 8. Impact of pre-treatments and sintering temperatures on (**A**) bending strength and (**B**) microhardness of sintered pellets.

During SPS sintering, graphite inlays in the sintering matrix can be a source of carbon diffusion into the Co pellet and may form a thin surface layer of carbide [36]. This can seriously impact the mechanical properties of the sintered pellet. To estimate the extent of carbon penetration, microhardness of the sintered pellet (set III, 1000 °C) was determined as a function of distance from the surface (Figure 9). Carbon penetration depth was estimated to be ~0.3 mm as indicated by the regions of low microhardness.

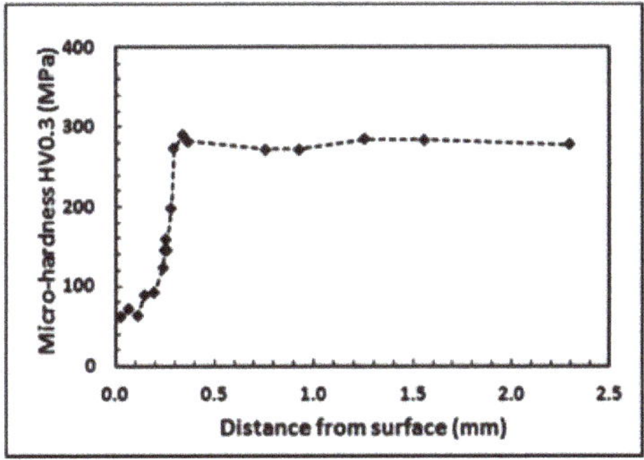

Figure 9. Impact of surface carbon diffusion on the microhardness of sintered Co pellets (set III, 1000 °C).

3.4. Discussion

This study was focused on the roles of initial pre-treatments including high energy mechanical treatment and nanomodification, and SPS consolidation temperatures on the basic characteristics of sintered Co pellets. Initial pre-treatments had a strong influence on the morphology of Co powders prior to sintering. While untreated Co powders (set I) were well rounded, high energy mechanical treatment (set II) caused major particle deformations, disintegrations, and the generation of odd shaped particles in a range of sizes. The addition of 0.5 wt. % Co nanopowders (set III) provided fillers for closing small gaps during sintering. These treatments had a strong influence on the grain sizes of the sintered pellets. The operating temperatures of SPS consolidation (700–1000 °C) were significantly lower than the melting point of cobalt (1495 °C). As very limited grain growth is expected during low temperature SPS sintering, the large grain size of sintered Co pellets (set I; 700 °C) is consistent with large particle sizes of as received Co powders (Figures 3a and 5a). Intensive crushing and associated deformations during HEMT (set II) and the additional presence of nano-modifiers (set III) resulted in a high degree of compaction, filling of gaps, and enhanced sintering. As sintering was carried out at fairly low temperatures, grain growth was strongly inhibited. Fine grain structure was observed for both sets II and III (Figure 6b,c; 700 °C).

This study showed very high degree of compactions reaching up to 99.6% relative densities in sintered Co pellets. This was achieved for set III pellets at 700 °C, 50 MPa pressure, 50 °C/min heating rate, and 5 min of holding time. Results on untreated Co powders (set I) show that highest relative density of 99.3% was achieved at 800 °C. Fellah et al. [37] achieved maximum relative density (94.5%) for cobalt nanopowders with SPS technique at 500 °C, 50 MPa pressure, 100 °C/min heating rate. Cabibbo (2018) used ball milling and SPS sintering on 2 μm cobalt powders to achieve a maximum relative density of 95.8% [38]. Lee et al. [2] achieved a relative density of 81–95% for tungsten powders with SPS technique at 1600–1800 °C, 60–120 MPa pressure, 100 °C/min heating rate, holding times up to 60 min.

Highest relative density of 99.4% for set II with HEMT pre-treatment was observed at 700 °C, a reduction of sintering temperature by 100 °C as compared to set I. This result clearly shows that HEMT processing could be used to lower SPS temperature without compromising sinter quality. Other authors have used high energy ball milling as a pre-treatment prior to SPS processing. Marek et al. [24] could achieve relative densities of 88.8–97.8% on Co powders; and Kundu et al. [25] achieved a maximum relative density of 98% for Fe-9Cr alloys. Even better results were observed for set III involving both HEMT and nanomodification reaching the highest compaction of 99.6% at 700 °C. The addition of nano-additives is known to inhibit grain growth by reinforcing particles at grain boundaries and limiting grain-boundary movement [39,40]. Reduced porosity, fine grain structure, and a high degree of compaction can also lead to improved mechanical properties of the sintered product.

The temperature of SPS consolidation was found to have a strong influence of densities, bending strength, and microhardness of sintered pellets. Highest degree of compaction (99.6%), highest bending strength (1997 MPa), highest microhardness (305 MPa) were observed for nanomodified, HEMT processed Co powders sintered at 700 °C. This temperature is even lower than the half melting temperature of Co. A direct correlation was observed between these basic characteristics and the sintering temperature. With temperature increasing from 700 °C, the magnitudes of densities, bending strength, and microhardness of sintered pellets were found to decrease, reaching a minimum at 900 °C and then starting to increase at 1000 °C. This basic trend was observed for all three sets of Co specimens to a certain degree. This trend cannot be explained in terms of initial particle shapes, morphology, and grain sizes, which were significantly different for the three sets of Co specimens (Figures 5 and 6).

During SPS sintering of Ti-4.8 wt. %TiB_2 composites, Namini et al. [41] observed a linear increase in relative density from 93.90% to 99.88%, with sintering temperature increasing from 750 to 1200 °C, thereby indicating high rates of porosity removal. However, the relative density of composites showed a small reduction at 1350 °C which was attributed to the softening of the matrix at higher temperatures, the grain growth surpassing densification postponing the porosity removal.

Figure 10 shows the dependence of bending strength and microhardness on the relative densities of sintered pellets. Data was taken from three sets of cobalt samples sintered at 700, 800, 900, and 1000 °C. Both mechanical properties were found to increase linearly with increasing densities. A linear correlation was also observed between the bending strength and microhardness of sintered pellets. These results indicate a direct link between the sinter density and mechanical properties irrespective of the sintering temperature. The impact of SPS temperature on densification and mechanical properties needs to be understood clearly. High resolution microscopic investigations with much smaller temperature steps will be required for creating an extensive database towards developing a fundamental understanding, associated mechanisms, and for optimizing the overall process.

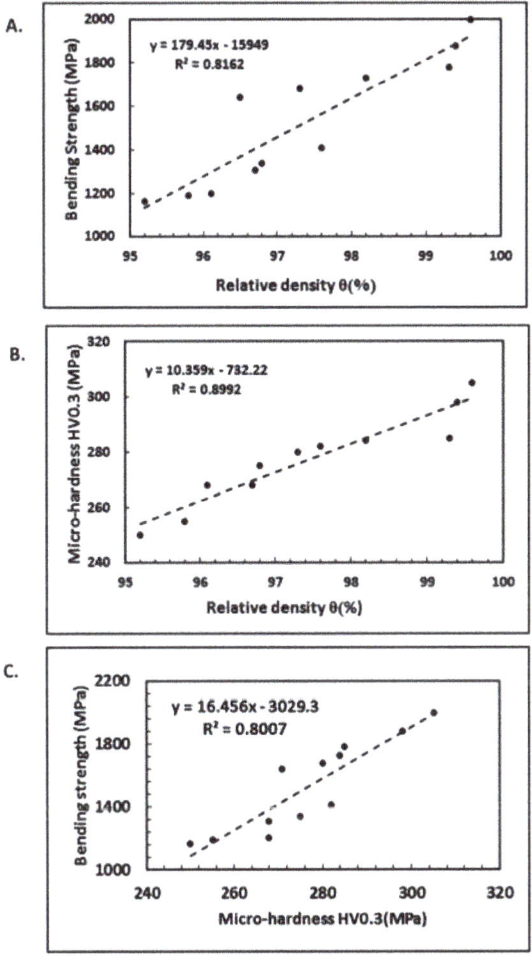

Figure 10. (**A–C**) Linear dependence of bending strength and microhardness on the relative density of sintered pellets.

4. Conclusions

Our results show that HEMT and nanomodification could be used to lower sintering temperatures while providing excellent results on basic characteristics of sintered products. Key findings of this study are:

(1). Three sets of Co micro-powders were investigated in this study: untreated, HEMT processed, and nanomodified HEMT processed powders. The SPS consolidation was carried out in the temperature range 700–1000 °C.

(2). The grain sizes of sintered Co pellets were largest for set I without any pre-treatment. Fine-grained products were obtained for Co powders undergoing HEMT processing and nanomodification. Nano-additives tended to inhibit grain growth by reinforcing particles at grain boundaries and limiting grain-boundary movement.

(3). Very high degrees of compaction were achieved with relative densities of sintered Co pellets, ranging from 95.2% to 99.6%. Pre-treatments as well as lower sintering temperatures were found to play a key role in enhancing the sinter quality. A reduction of sintering temperature by 100 °C was observed for pre-treated Co powders as compared to untreated Co powders for a comparable sinter quality.

(4). Direct and linear co-relations were observed between the mechanical properties and densities of sintered Co pellets. Highest values of bending strength (1997 MPa), microhardness (305 MPa), and relative density (99.6%) were observed for nanomodified HEMT and SPS processed Co pellets sintered at 700 °C.

(5). Lowest values of bending strength (1198 MPa), microhardness (268 MPa), and relative density (96.1%) were observed for nanomodified HEMT and SPS processed Co pellets sintered at 900 °C.

(6). This study has shown that SPS operating temperatures need to be carefully optimized to prevent inadequate sintering at low temperatures and increased porosity, defect formation, and product degradation at high temperatures.

Author Contributions: Conceptualization, V.M.N., Y.K., G.K.; Methodology, V.M.N., T.H.N., Y.K.; Investigation, V.M.N., I.B., V.L., R.K., Writing and Editing, Y.K., V.M.N., I.G., R.K. All authors have read and agreed to the published version of the manuscript.

Funding: This research received no external funding.

Conflicts of Interest: The authors declare no conflict of interest.

References

1. Matizamhuka, W.R. Spark plasma sintering (SPS)—An advanced sintering technique for structural nanocomposite materials. *J. S. Afr. Inst. Min. Metall.* **2016**, *116*, 1171–1180. [CrossRef]
2. Lee, G.; McKittrick, J.; Ivanov, E.; Olevsky, E.A. Densification mechanism and mechanical properties of tungsten powder consolidated by spark plasma sintering. *Int. J. Refr. Met. Hard Mater.* **2017**, *61*, 22–29. [CrossRef]
3. Guillon, O.; Gonzalez-Julian, J.; Dargatz, B.; Kessel, T.; Schierning, G.; Rathel, J.J.; Herrmann, M. Field-assisted sintering technology/spark plasma sintering: Mechanisms, Materials, and Technology Developments. *Adv. Eng. Mater.* **2014**, *16*, 830–849. [CrossRef]
4. Suárez, M.; Fernández, A.; Menéndez, J.L.; Torrecillas, R.; Kessel, H.U.; Hennicke, J.; Kirchner, R.; Kessel, T. Challenges and opportunities for spark plasma sintering: A key technology for a new generation of materials. *Sinter. Appl.* **2013**, *13*, 319–342.
5. Munir, Z.A.; Anselmi-Tamburini, U.M.; Ohyanagi, M. The effect of electric field and pressure on the synthesis and consolidation of materials: A review of the spark plasma sintering method. *J. Mater. Sci.* **2006**, *41*, 763–777. [CrossRef]
6. Orrù, R.; Richeri, R.; Locci, A.M.; Cincotti, A.; Cao, G. Consolidation/synthesis of materials by electric current activated/assisted sintering. *Mater. Sci. Eng. R* **2009**, *63*, 127–287. [CrossRef]
7. Mamedov, V. Spark plasma sintering as advanced PM sintering method. *Powder Metall.* **2002**, *45*, 322–328. [CrossRef]
8. Huang, J.L.; Nayak, P.K. Strengthening alumina ceramic matrix nanocomposites using spark plasma sintering. In *Advances in Ceramic Matrix Composites*, 2nd ed.; Elsevier Inc.: Amsterdam, The Netherlands, 2018; pp. 231–247. [CrossRef]

9. Xie, J. Spark Plasma Sintering: A useful technique to develop large-sized bulk metallic glasses. *J. Powder Metall. Min.* **2013**, *2*, e109. [CrossRef]
10. Xie, G.Q.; Qin, F.X.; Zhu, S.L. Recent progress in Ti-based metallic glasses for application as biomaterials. *Mater. Trans.* **2013**, *54*, 1314–1323. [CrossRef]
11. Mellor, I.; Doughty, G. Novel and emerging routes for titanium powder production—An overview. *Key Eng. Mater.* **2016**, *704*, 271–281. [CrossRef]
12. Méar, F.O.; Xie, G.Q.; Louzguine-Luzgin, D.V.; Inoue, A. Spark plasma sintering of Mg-based amorphous ball-milled powders. *Mater. Trans.* **2009**, *50*, 588–591.
13. Zheng, B.L.; Ashford, D.; Zhou, Y.Z.; Mathaudhu, S.N.; Delplanque, J.P. Influence of mechanically milled powder and high pressure on spark plasma sintering of Mg-Cu-Gd metallic glasses. *Acta Mater.* **2013**, *6*, 4414–4428. [CrossRef]
14. Dudina, D.V.; Bokhonov, B.B.; Olevsky, E.A. Fabrication of porous materials by spark plasma sintering: A review. *Materials* **2019**, *12*, 541. [CrossRef] [PubMed]
15. Vincent, N. Spark Plasma Sintering of Titanium and Cobalt Alloys for Biomedical Applications. Ph.D. Thesis, University Trento, Trento, Italy, 2012.
16. Tang, C.F.; Pan, F.; Qu, X.H.; Jia, C.C.; Duan, B.H.; He, X.B. Spark plasma sintering cobalt base superalloy strengthened by Y–Cr–O compound through high-energy milling. *J. Mater. Process. Technol.* **2008**, *204*, 111–116. [CrossRef]
17. Gezerman, A.O.; ÇorbacJoLlu, B.D. Effects of Mechanical Alloying on Sintering Behavior of Tungsten Carbide-Cobalt Hard Metal System. *Adv. Mater. Sci. Eng.* **2017**, *2017*, 8175034. [CrossRef]
18. Patel, B.; Favaro, G.; Inam, F.; Reece, M.J.; Angadji, A.; Bonfield, W.; Huang, J.; Edirisinghe, M. Cobalt-based orthopaedic alloys: Relationship between forming route, microstructure and tribological performance. *Mater. Sci. Eng. C* **2012**, *32*, 1222–1229. [CrossRef]
19. Cavaliere, P.; Sadeghi, B.; Shabani, A. Spark plasma sintering: Process fundamentals. In *Spark Plasma Sintering of Materials*; Cavaliere, P., Ed.; Springer Nature Switzerland: Cham, Switzerland, 2019. [CrossRef]
20. Song, X.; Liu, X.; Zhang, J. Neck formation and self-adjusting mechanism of neck growth of conducting powders in spark plasma sintering. *J. Am. Ceram. Soc.* **2006**, *89*, 494–500. [CrossRef]
21. Oke, S.R.; Ige, O.O.; Falodun, O.E.; Obadele, B.A.; Shongwe, M.B.; Olubambi, P.A. Optimization of process parameters for spark plasma sintering of nano structured SAF 2205 composite. *J. Mater. Res. Technol.* **2018**, *7*, 126–134. [CrossRef]
22. Deng, S.; Li, R.; Yuan, T.; Cao, P.; Xie, S. Electromigration-enhanced densification kinetics during spark plasma sintering of tungsten powder. *Metall. Mater. Trans. A* **2019**, *50*, 2886–2897. [CrossRef]
23. Knaislová, A.; Novák, P.; Cygan, S.; Jaworska, L.; Cabibbo, M. High-pressure spark plasma sintering (HP SPS): A promising and reliable method for preparing Ti–Al–Si alloys. *Materials* **2017**, *10*, 465. [CrossRef]
24. Marek, I.; Vojtěch, D.; Michalcová, A.; Kubatík, T.F. The structure and mechanical properties of high-strength bulk ultrafine-grained Cobalt prepared using high-energy ball milling in combination with spark plasma sintering. *Materials* **2016**, *9*, 391. [CrossRef] [PubMed]
25. Kundu, A.; Sittiho, A.; Charit, I.; Jaques, B.; Jiang, C. Development of Fe-9Cr alloy via high-energy ball milling and spark plasma sintering. *JOM* **2019**, *71*, 2846–2855. [CrossRef]
26. Coutsouradis, D.; Davin, A.; Lamberigts, M. Cobalt-based superalloys for applications in gas turbines. *Mater. Sci. Eng.* **1987**, *88*, 11–19. [CrossRef]
27. Suzuki, A.; Inui, H.; Pollock, T.M. L12-strengthened cobalt-base superalloys. *Annu. Rev. Mater. Res.* **2015**, *45*, 345–368. [CrossRef]
28. Aherwar, A.; Singh, A.; Patnaik, A. Cobalt based alloy: A better choice biomaterial for hip implants. *Trends Biomater. Artif. Organs* **2016**, *30*, 50–55.
29. Prakasam, M.; Locs, J.; Salma-Ancane, K.; Loca, D.; Largeteau, A.; Berzina-Cimdina, L. Biodegradable materials and metallic implants—A review. *J. Funct. Biomater.* **2017**, *8*, 44. [CrossRef]
30. Van Minh, N.; Konyukhov, Y.; Karunakaran, G.; Ryzhonkov, D. Enhancement of densification and sintering behaviour of tungsten material via nano modification and magnetic mixing processed under spark plasma sintering. *Metals Mater. Int.* **2017**, *23*, 532–542. [CrossRef]
31. Van Minh, N.; Karunakaran, G.; Konyukhov, Y. Effect of mixing modes and nano additives on the densification and sintering behavior of tungsten material under spark plasma sintering. *J. Clust. Sci.* **2017**, *28*, 2157–2165. [CrossRef]

32. Tiryakioğlu, M.; Dispinar, D.; Uludağ, M.; Yazman, S.; Gemi, L. The effect of 0.5 wt.% additions of carbon nanotubes & ceramic nanoparticles on tensile properties of epoxy-matrix composites: A comparative study. *Mater. Sci. Nanotechnol.* **2017**, *1*, 15–22.
33. Xing, W.Q.; Yu, X.Y.; Li, H.; Ma, L.; Zuo, W.; Dong, P.; Wang, W.X.; Ding, M. Effect of nano Al_2O_3 additions on the interfacial behaviour and mechanical properties of eutectic Sn-9Zn solder on low temperature wetting and soldering of 6061 aluminium alloys. *J. Alloys Compd.* **2017**, *695*, 574–582. [CrossRef]
34. Gain, A.K.; Zhang, L. Effects of Ni nanoparticles addition on the microstructure, electrical and mechanical properties of Sn-Ag-Cu alloy. *Materialia* **2019**, *5*, 100234. [CrossRef]
35. Gain, A.K.; Zhang, L. Interfacial microstructure, wettability and material properties of nickel (Ni) nanoparticle doped tin-bismuth-silver (Sn-Bi-Ag) solder on copper (Cu) substrate. *J. Mater. Sci. Mater. Electron.* **2016**, *27*, 1–13. [CrossRef]
36. Bokhonov, B.B.; Ukhina, A.V.; Dudina, D.V.; Anisimov, A.G.; Mali, V.I.; Batraev, I.S. Carbon uptake during Spark Plasma Sintering: Investigation through the analysis of the carbide "footprint" in a Ni-W alloy. *RSC Adv.* **2015**, *5*, 80228–80237. [CrossRef]
37. Fellah, F.; Schoenstein, F.; Dakhlaoui–Omrani, A.; Chérif, S.M.; Dirras, G.; Jouini, N. Nanostructured cobalt powders synthesised by polyol process and consolidated by Spark Plasma Sintering: Microstructure and mechanical properties. *Mater. Charact.* **2012**, *69*, 1–8. [CrossRef]
38. Cabibbo, M. Nanostructured cobalt obtained by combining bottom-up and top-down approach. *Metals* **2018**, *8*, 962. [CrossRef]
39. Rogal, L.; Kalita, D.; Tarasek, A.; Bobrowski, P.; Czerwinski, F. Effect of SiC nano-particles on microstructure and mechanical properties of the CoCrFeMnNi high entropy alloy. *J. Alloys Compd.* **2017**, *708*, 344–352. [CrossRef]
40. Karunakaran, G.; Van Minh, N.; Konyukhov, Y.; Kolesnikov, E.; Venkatesh, M.; Kumar, G.S.; Gusev, A.; Kuznetsov, D. Effect of Si, B, Al_2O_3 and ZrO_2 nano-modifiers on the structural and mechanical properties of Fe + 0.5 % C alloy. *Arch. Civ. Mech. Eng.* **2017**, *17*, 669–676. [CrossRef]
41. Namini, A.B.; Motallebzadeh, A.; Nayebi, B.; Mehdi Shahedi, A.M.; Azadbeh, M. Microstructure–mechanical properties correlation in spark plasma sintered Ti–4.8 wt.% TiB2 composites. *Mater. Chem. Phys.* **2018**, *223*, 789–796. [CrossRef]

Article

Photochemical Synthesis of Silver Nanodecahedrons under Blue LED Irradiation and Their SERS Activity

Mai Ngoc Tuan Anh [1,2], Dinh Tien Dung Nguyen [3], Ngo Vo Ke Thanh [1], Nguyen Thi Phuong Phong [2], Dai Hai Nguyen [4,5] and Minh-Tri Nguyen-Le [6,7,*]

1 Research Laboratories of Sai Gon Hi-Tech Park, Lot I3, N2 Street, Tan Phu Ward, District 9, Ho Chi Minh City 70000, Vietnam; anhmnt.shtplabs@gmail.com (M.N.T.A.); nvkthanh.shtp@tphcm.gov.vn (N.V.K.T.)

2 Faculty of Chemistry, University of Science, Vietnam National University Ho Chi Minh City, 227 Nguyen Van Cu Street, District 5, Ho Chi Minh City 70000, Vietnam; ntpphong@hcmus.edu.vn

3 Institute of Research and Development, Duy Tan University, Danang 550000, Vietnam; nguyendtiendung@duytan.edu.vn

4 Institute of Applied Materials Science, Vietnam Academy of Science and Technology, 01 TL29, District 12, Ho Chi Minh City 700000, Vietnam; nguyendaihai0511@gmail.com

5 Graduate University of Science and Technology, Vietnam Academy of Science and Technology, 18 Hoang Quoc Viet, Cau Giay, Hanoi 100000, Vietnam

6 Laboratory of Advanced Materials Chemistry, Advanced Institute of Materials Science, Ton Duc Thang University, Ho Chi Minh City 758307, Vietnam

7 Faculty of Applied Sciences, Ton Duc Thang University, Ho Chi Minh City 758307, Vietnam

* Correspondence: nguyenleminhtri@tdtu.edu.vn

Received: 15 February 2020; Accepted: 2 March 2020; Published: 4 March 2020

Abstract: Silver nanodecahedrons were successfully synthesized by a photochemical method under irradiation of blue light-emitting diodes (LEDs). The formation of silver nanodecahedrons at different LED irradiation times (0–72 h) was thoroughly investigated by employing different characterization methods such as ultraviolet–visible spectroscopy (UV–Vis), transmission electron microscopy (TEM), and Raman spectroscopy. The results showed that silver nanodecahedrons (AgNDs) were formed from silver nanoseeds after 6 h of LED irradiation. The surface-enhanced Raman scattering (SERS) effects of the synthesized AgNDs were also studied in comparison with those of spherical silver nanoparticles in the detection of 4-mercapto benzoic acid. Silver nanodecahedrons with a size of 48 nm formed after 48 h of LED irradiation displayed stronger SERS properties than spherical nanoparticles because of electromagnetic enhancement. The formation mechanism of silver nanodecahedrons is also reported in our study. The results showed that multihedral silver nanoseeds favored the formation of silver nanodecahedrons.

Keywords: silver nanodecahedron; SERS; photochemical synthesis; LEDs

1. Introduction

Recently, silver nanomaterials have gained much attention due to their various applications in the areas of electronic and sensor engineering, catalysis, biomedicine, and surface-enhanced Raman scattering (SERS) [1,2]. For the most part, these applications are based on unique optical properties of silver nanomaterials, particularly their light absorption and scattering ability as a result of surface plasmon resonance [1,3]. The optical properties depend on the shape and size of the nanoparticles, therefore in the last decade, many groups all over the world have studied the synthesis of silver nanomaterials with shape- and size-controlled anisotropy, such as plate-, cube-, wire-, rod-, pyramid-, and decahedron-shaped silver [1–4]. Among these, silver nanoplates (AgNPs) and silver nanodecahedrons (AgNDs) have been studied because of their localized surface plasmon resonance in a longer wavelength region. In particular,

AgNDs with multihedral structure display a remarkable scattering ability [3,4]. So far, AgNDs have been synthesized by a chemical method using hydrazine as a reductive agent and trisodium citrate as a stabilizer [5]. However, only few AgNDs are obtained with this method. The synthesis of AgNDs using a photochemical method based on irradiation by light-emitting diodes (LEDs) for a better control of their shape has gained much attention [3,4,6–10]. One of the first researchers to study AgNDs were Stamplecoskie and Scaiano. In their study, AgNDs were synthesized via two steps: firstly, the synthesis of 3 nm spherical AgNPs with the optical reductive agent I-2959 and, secondly, their irradiation using 455 nm blue LEDs. However, the reaction mechanism is not known [6]. Jamil Saade et al. also synthesized AgNDs by irradiating 3–5 nm AgNP seeds with blue LEDs, but no applications of AgNDs were mentioned [7]. Shan-Wei Lee et al. first reported the synthesis of AgNDs based on the two aforementioned steps but using 520 nm green LEDs at low temperature for SERS applications. Although these authors claimed that the obtained AgNDs have better SERS properties than AgNPs, no persuasive explanation was given [8]. While studying the thermodynamics of formation of AgNDs, Haitao Wang et al. realized that using LEDs at a low temperature would facilitate the formation of AgNDs. They studied the mechanism of AgNDs formation but did not consider the surface plasmon effect, one of the important surface properties of silver nanomaterials [9]. Cardoso-Avila et al. also applied LED irradiation on quartz cuvettes containing AgNPs to synthesize AgNDs; however, neither the reaction mechanism nor the SERS effects were thoroughly investigated [10,11].

In this study, we synthesized AgNDs under blue LED irradiation via a two-step photochemical method. The formation of AgNDs was investigated at different LED irradiation times. The SERS effects of the synthesized AgNDs were also studied in comparison with those of AgNPs in the detection of 4-mercapto benzoic acid (4-MBA), a Raman tester able to bind to silver nanoparticles through Ag–S bonds, enhancing the SERS effects [12–14].

2. Materials and Methods

2.1. Chemicals

Silver nitrate ($AgNO_3$, >99%, Sigma-Aldrich, Darmstadt, Germany), trisodium citrate tribasic dihydrate (TSC, >99%, Sigma-Aldrich, Darmstadt, Germany), polyvinylpyrrolidone (PVP K30, Prolabo, Kennersburg, NJ, USA), L-arginine (L-A, 99%, Merck, Darmstadt, Germany), sodium borohydride ($NaBH_4$, 98%, Merck, Darmstadt, Germany), 4-mercapto benzoic acid (4-MBA, 99%, Sigma-Aldrich, Darmstadt, Germany), and DI water (standard HPLC, Merck, Darmstadt, Germany). Chemicals were used without any purification.

2.2. Synthesis of Silver Nanodecahedrons

AgNDs were synthesized in 2 steps: (i) synthesis of AgNPs as seeds and (ii) irradiating the seeds with blue LEDs to grow AgNDs with the aid of L-A [3,11]. The synthesis procedure is described in Figure 1.

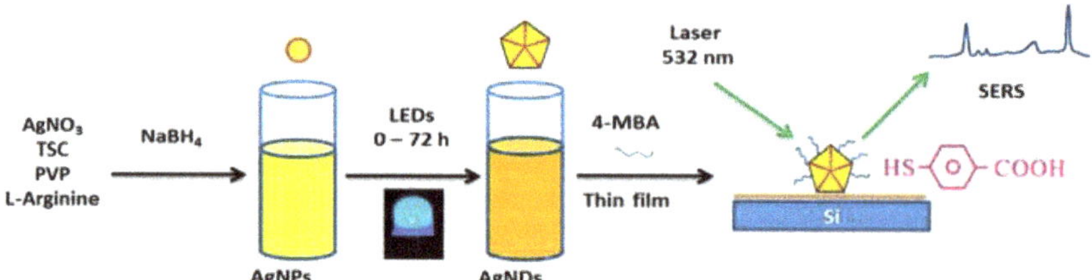

Figure 1. Schematic illustration of the fabrication and surface-enhanced Raman scattering (SERS) measurement of nanodecahedrons (AgNDs). TSC: trisodium citrate tribasic dihydrate, PVP: polyvinylpyrrolidone, AgNPs: silver nanoplates, LEDs: light-emitting diodes, 4-MBA: 4-mercapto benzoic acid.

2.3. Synthesis of Seeds

All glass equipment and magnetic stirrers were treated with aqua regia before use. In total, 5 mL of 50 mM TSC, 150 μL of 50 mM PVP, 2 mL of 5 mM $AgNO_3$, and 250 μL of 5 mM L-A were mixed in a beaker, then milli-Q water was added into the mixture to 100 mL. The mixture was magnetically stirred at 500 rpm for 5 min. Afterwards, 800 μL of 100 mM $NaBH_4$ was rapidly added to the solution. The color of the solution immediately changed into light yellow, then yellow, indicating the presence of AgNPs in the solution. The solution was stirred at room temperature and aged in the dark overnight.

2.4. Synthesis of AgNDs

A volume of 20 mL of the as-synthesized AgNPs solution stored in a glass bottle (Wheaton-Germany) was exposed to blue LEDs (Dragon—Taiwan, 460 ± 12 nm, output: 10 W) for 6, 12, 24, 48, and 72 h; the color of the solution changed from yellow to orange.

2.5. Characterization

The morphological structure of the synthesized AgNDs was studied through a JEOL JEM-1400 (Jeol Ltd., Tokyo, Japan). The samples were dropped onto 3 mm-diameter copper grids, dried at room temperature, and analyzed by TEM (Jeol Ltd., Tokyo, Japan) at 100 kV. The light absorption spectra of AgNDs and intermediates were analyzed by a V-750 UV/Vis spectrophotometer (Jasco Co., Tokyo, Japan, 100 nm scanning speed per minute).

2.6. SERS Measurements

SERS in the presence of the reductive agent 4-MBA by silver nano-seeds and AgNDs obtained after 48 h of LED irradiation was measured on a Si wafer. The wafer was cut into 1 × 1 cm slices and treated with piranha solution to remove all organic compounds. Then, 2 mL of AgNDs (or AgNPs) solution was centrifuged at 12,000 rpm for 15 min to obtain a precipitate. The precipitate was then redispersed into 1 mL of DI water; 50 μL of 10^{-5} M 4-MBA solution was mixed with 450 μL of the sample solution for 1 h at room temperature. Afterwards, 20 μL of the obtained mixture was dropped on the Si wafer and dried at room temperature, obtaining a 4-MBA@AgNDs (or AgNPs) @Si wafer ready for Raman analysis, carried out with HORIBA XploRA ONE TM, Palaiseau, France, equipped with a 532 nm laser source and a 10x optical microscope lens. The sample was analyzed at 10 random locations to obtain an average value. For comparison, the Raman spectrum of a 10^{-4} M 4-MBA solution without AgNDs or AgNPs was also recorded with the same procedure.

3. Results

3.1. Synthesis of Silver Nanodecahedrons

The UV–Vis results in Figure 2 show the formation of silver nanoseeds (at 0 h) which presented an absorption peak at 400 nm (peak 1). A yellow color indicated the presence of spherical silver nanoparticles in the solution [4–9]. After irradiating with LEDs for 6 h, this peak showed a red shift to 407 nm, and another peak (peak 2) appeared at 460 nm, corresponding to the vibration of dipole resonance of AgNDs [6–9]. It was concluded that AgNDs were formed after 6 h of LED irradiation. After increasing the irradiation time to 12 h, the UV–Vis spectra showed red shifts of both peak 1 and peak 2 toward higher wavelengths, i.e., 413 nm and 470 nm, respectively. On the other hand, the absorbance of peak 1 decreased, while that of peak 2 increased rapidly due to the generation of more AgNDs from AgNPs after 12 h of irradiation. After 24 h, peak 1 was shifted toward 404 nm, and the absorbance decreased from 1.1 to 0.5 a.u., proving the number of AgNPs in the solution continued to decrease under LED irradiation. Peak 2 showed a decrease in the absorbance from 2.0 to 1.3, but the wavelength did not change much. On the other hand, after 24 h, peak 3 appeared at 345 nm, in accordance with AgNDs' out-of-plane quadrupole resonance [6,9]. When we continued to increase the irradiating time to 48 and 72 h, we did not observe significant changes in the UV–Vis spectra. In fact, compared to the 24 h spectrum, peak 1 and peak 3 intensity almost did not change, while the intensity of peak 2 (464 nm), corresponding to AgNDs's dipole resonance, increased to 1.81. In conclusion, 48 h of irradiation was a suitable time to transform AgNPs to AgNDs.

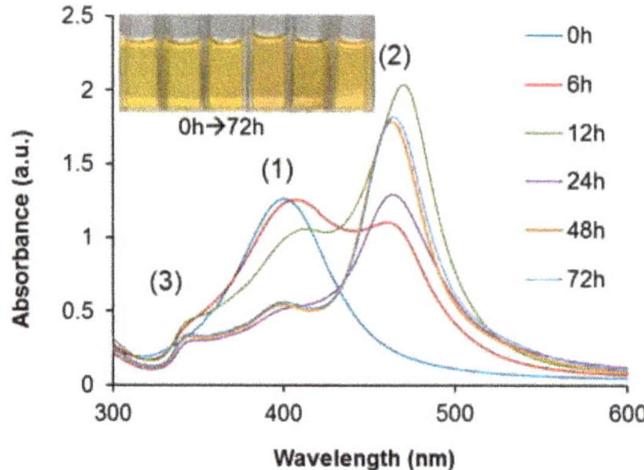

Figure 2. UV–Vis spectra of silver nanoparticles at different LED irradiation times.

The TEM images in Figure 3a revealed that AgNDs with average size of 30 nm and AgNPs smaller than 10 nm on average were the dominant species in the sample irradiated for 6 h. After 48 h of irradiation, the TEM image showed the presence in the sample of AgNDs with average size of 48 nm, in addition to AgNPs smaller than 10 nm on average. Moreover, the TEM results also indicated that increasing the LED irradiation time would result in the combination of AgNPs, leading to the formation of AgNDs, which was consistent with the UV–Vis results.

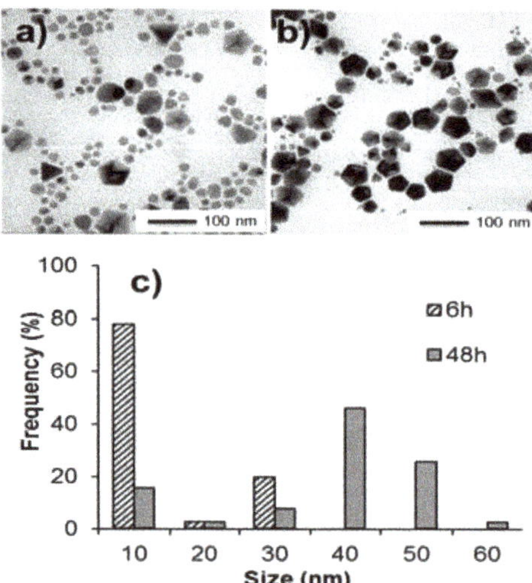

Figure 3. (**a**,**b**) TEM images and (**c**) particle size distribution of silver nanoparticles after 6 h and 48 h of irradiation.

The formation mechanism of AgNDs under blue LEDs is shown in Figure 4. Due to their surface plasmon resonance properties, the silver nanoparticles were able to combine to form anisotropic silver nanoparticles with larger size when they were irradiated by LEDs, which provided enough energy for surface plasmon resonance [9]. The combination is illustrated in Figure 4. We propose two plausible growth processes for the formation of Ag decahedrons.

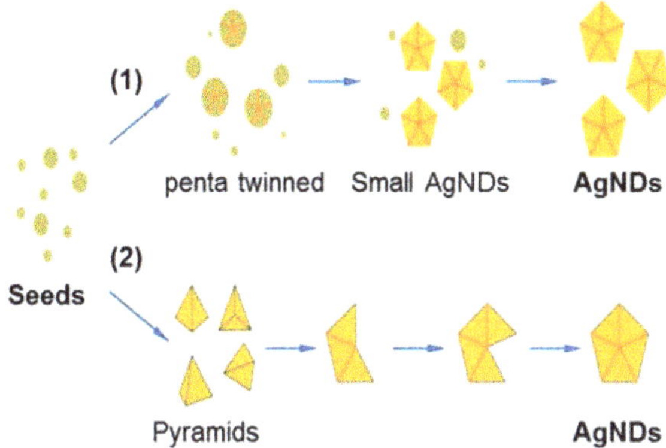

Figure 4. Proposed mechanisms of formation of AgNDs.

Mechanism 1: AgNP seeds were first generated as multihedral seeds with a small size, then combined with new-born seeds to form larger AgNDs [3,4,7,9]. It is widely accepted that the shape of multihedral seeds strongly depends on the excitation wavelength. For example, irradiation with blue LEDs would result in the formation of penta-twinned AgNDs [4,7,9,15], while planar twinned seeds can be generated under green LEDs [6–8]. The TEM results of AgNDs formed after 6 h of LED irradiation in this study showed the existence of AgNDs with sizes smaller than 20 nm, supporting this mechanism.

Mechanism 2: Silver nanopyramids were first generated due to the combination of AgNPs. Five pyramids were then combined along their side edges to form AgNDs during LED irradiation [4,7]. The TEM results of AgNDs observed after 48 h of irradiation (Figure 3) showed the presence of some silver nanopyramids and incomplete AgNDs, providing a proof of this mechanism.

In general, both mechanisms are possible during LED irradiation. Compared to mechanism 2, mechanism 1 is favored because of the low probability of forming AgNDs from five pyramid nanoparticles. Moreover, according to the UV–Vis results, the dramatic decrease in intensity of both peak 1 and 2 after 24 h of LED irradiation, as well as the gradual increase in intensity of peak 2 and the negligible change in intensity of peak 1 as the LED irradiation time of the silver solution kept increasing up to 48 h, indicated the aggregation of small AgNDs into larger AgNDs, which further supports mechanism 1.

3.2. SERS Properties of AgNDs

The Raman spectra of 4-MBA powder shown in Figure 5c revealed two peaks at 1091 and 1597 cm^{-1}, corresponding to the aromatic ring vibration, and two peaks at 1150 and 1180 cm^{-1}, corresponding to the COO^- vibration [12–14]. The SERS spectra of the mixture of 10^{-5} M 4-MBA and AgNPs or AgNDs showed that the peaks of the aromatic ring were shifted toward 1082 and 1589 cm^{-1} (Figure 5a,b). The remarkably higher intensity of the Raman peaks compared with the flat Raman signal of 10^{-4} M 4-MBA (as well as of 10^{-5} M 4-MBA) (Figure 5d) indicated that both materials were capable of enhancing SERS; in particular, AgNDs showed better SERS enhancement than AgNPs.

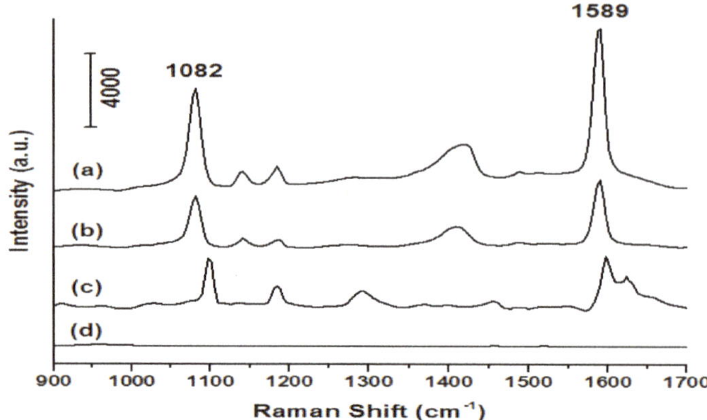

Figure 5. Raman spectra of the mixture of 4-MBA (10^{-5} M) and (**a**) AgNDs (after 48 h of LED irradiation) or (**b**) AgNPs; Raman spectra of (**c**) 4-MBA powder and (**d**) 4-MBA (10^{-4} M).

The enhancement of SERS was probably due to chemical and electromagnetic mechanisms [12,13]. The chemical mechanism depends on the binding of 4-MBA and nano-silver. Since both AgNPs and AgNDs were capable of binding 4-MBA through Ag–S bonds, the difference in SERS properties was due to the electronic mechanism, which is deeply related to the surface plasmon properties of nano-silver. AgNDs have better surface plasmon properties than AgNPs, as displayed by the UV–Vis results, showing a bipolar peak for AgNPs twice as big as that of AgNDs. On the other hand, because the excited wavelength was 532 nm, closer to the wavelength of the bipolar peak of AgNPs (464 nm) than to that of the peak of AgNDs (400 nm), AgNDs were capable of better enhancing SERS, in agreement with Haifei Lu's report [3]. Besides, energy-concentrating spots called "hot spot" formed on AgNDs' surface, which were located at the points and lines of the decahedrons [3,4,12,13] corresponding to face (111). These spots further enhanced the SERS properties of AgNDs.

4. Conclusions

We successfully synthesized silver decahedrons nanomaterials (AgNDs) by a photochemical method using blue 10 W LEDs. After 48 h of LED irradiation, silver nanoparticles with size smaller than 10 nm combined with each other, forming AgNDs with average size of 48 mn, which were analyzed by UV–Vis spectroscopy and TEM. The results indicated that AgNDs were formed through intermediates such as small AgNDs seeds or silver nanopyramids. Both silver nanoseeds and AgNDs exhibited good SERS enhancement in the presence of 4-MBA. AgNDs had better SERS effects than AgNPs, which can be explained by an electronic mechanism.

Author Contributions: Conceptualization, N.T.P.P. and D.H.N. methodology, N.T.P.P. and D.H.N. formal analysis, M.N.T.A., D.T.D.N., and N.V.K.T. investigation, M.N.T.A. and D.T.D.N. writing—original draft preparation, M.N.T.A. and M.-T.N.-L. writing—review and editing, M.-T.N.-L. supervision, M.-T.N.-L. administration, D.H.N. funding acquisition, D.H.N. and M.-T.N.-L. All authors have read and agreed to the published version of the manuscript.

Funding: This study was funded by the People's Committee of Ho Chi Minh City, performed with the support of the Board of Management of Saigon High-Tech Park (SHTP), the Research Laboratory of Saigon Hi-Tech Park (SHTPLABs), and the Institute of Applied Materials Science, Vietnam Academy of Science and Technology (IAMS).

Conflicts of Interest: The authors declare no conflict of interest.

References

1. Rycenga, M.; Cobley, C.M.; Zeng, J.; Li, W.; Moran, C.H.; Zhang, Q.; Qin, D.; Xia, Y. Controlling the Synthesis and Assembly of Silver Nanostructures for Plasmonic Applications. *Chem. Rev.* **2011**, *111*, 3669–3712. [CrossRef] [PubMed]

2. Zhou, S.; Zhao, M.; Yang, T.H.; Xia, Y. Decahedral nanocrystals of noble metals: Synthesis, characterization, and applications. *Mater. Today* **2017**, *22*, 108–131. [CrossRef]
3. Lu, H.; Zhang, H.; Yu, X.; Zeng, S.; Yong, K.T.; Ho, H. Seed-mediated Plasmon-driven Regrowth of Silver Nanodecahedrons (NDs). *Plasmonics* **2012**, *7*, 167–173. [CrossRef]
4. Zheng, X.; Zhao, X.; Guo, D.; Tang, B.; Xu, S.; Zhao, B.; Xu, W. Photochemical Formation of Silver Nanodecahedra: Structural Selection by the Excitation Wavelength. *Langmuir* **2009**, *25*, 3802–3807. [CrossRef] [PubMed]
5. Dadhich, B.K.; Kumar, I.; Choubey, R.K.; Bhushan, B.; Priya, A. Shape and size dependent nonlinear refraction and absorption in citrate-stabilized, near-IR plasmonic silver nanopyramids. *Photochem. Photobiol. Sci.* **2017**, *16*, 1556–1562. [CrossRef] [PubMed]
6. Stamplecoskie, K.G.; Scaiano, J.C. Light Emitting Diode Irradiation Can Control the Morphology and Optical Properties of Silver Nano particles. *JACS* **2010**, *132*, 1825–1827. [CrossRef] [PubMed]
7. Saade, J.; de Araújo, C.B. Synthesis of silver nanoprisms: A photochemical approach using light emission diodes. *Mater. Chem. Phys.* **2014**, *148*, 1184–1193. [CrossRef]
8. Lee, S.W.; Chang, S.H.; Lai, Y.S.; Lin, C.C.; Tsai, C.C.; Lee, Y.C.; Chen, J.C.; Huang, C.L. Effect of Temperature on the Growth of Silver Nano particles Using Plasmon-Mediated Method under the Irradiation of Green LEDs. *Materials* **2014**, *7*, 7781–7798. [CrossRef] [PubMed]
9. Wang, H.; Cui, X.; Guan, W.; Zheng, X.; Zhao, H.; Wang, Z.; Wang, Q.; Xue, T.; Liu, C.; Singh, D.J.; et al. Kinetic effects in the photomediated synthesis of silver nanodecahedra and nanoprisms: Combined effect of wavelength and temperature. *Nanoscale* **2014**, *6*, 7295–7302. [CrossRef] [PubMed]
10. Cardoso-Avila, P.E.; Molina, J.L.P.; Krishna, C.M.; Castro-Beltran, R. Photochemical transformation of silver nano particles by combining blue and green irradiation. *J. Nanopart. Res.* **2015**, *17*, 160. [CrossRef]
11. Cardoso-Avila, P.E.; Molina, J.L.P.; Kumar, K.U.; Arenas-Alatorre, J.A. Temperature and amino acid-assisted size- and morphology-controlled photochemical synthesis of silver decahedral nano particles. *J. Exp. Nanosci.* **2014**, *9*, 639–651. [CrossRef]
12. Weng, G.; Feng, Y.; Zhao, J.; Li, J.; Zhu, J.; Zhao, J. Size dependent SERS activity of Ag triangular nanoplates on different substrates: Glass vs. paper. *Appl. Surf. Sci.* **2019**, *478*, 275–283. [CrossRef]
13. Tran, D.M.; Lee, B.K.; Nguyen, L.M.T. Methanol-dispersed of ternary Fe_3O_4@γ-APS/graphene oxide-based nanohybrid for novel removal of benzotriazole from aqueous solution. *J. Environ. Manag.* **2018**, *209*, 452–461.
14. Lai, W.; Zhou, J.; Liu, Y.; Jia, Z.; Xie, S.; Petti, L.; Mormile, P. 4MBA-labeled Ag-nanorod aggregates coated with SiO2: Synthesis, SERS activity, and biosensing applications. *Anal. Methods* **2015**, *7*, 8832–8838. [CrossRef]
15. Elechiguerra, J.L.; Reyes-Gasgab, J.; Yacaman, M.J. The role of twinning in shape evolution of anisotropic noble metal nanostructures. *J. Mater. Chem.* **2006**, *16*, 3906–3919. [CrossRef]

Article

Maleated Natural Rubber/Halloysite Nanotubes Composites

Nabil Hayeemasae [1,*], Zareedan Sensem [1], Kannika Sahakaro [1] and Hanafi Ismail [2]

[1] Department of Rubber Technology and Polymer Science, Faculty of Science and Technology, Prince of Songkla University, Pattani Campus, Pattani 94000, Thailand; zareedan-4579@hotmail.com (Z.S.); kannika.sah@psu.ac.th (K.S.)

[2] School of Materials and Mineral Resources Engineering, Engineering Campus, Universiti Sains Malaysia, Nibong Tebal 14300, Malaysia; ihanafi@usm.my

* Correspondence: nabil.h@psu.ac.th

Received: 24 January 2020; Accepted: 25 February 2020; Published: 3 March 2020

Abstract: In this study, maleic anhydride (MA) grafted natural rubber (NR), known as maleated natural rubber (MNR), was melt-prepared with the MA content varied within 1–8 phr. MNR was used as the main matrix, with Halloysite Nanotubes (HNT) as a filler, in order to obtain composites with improved performance. The compounds were investigated for their filler–filler interactions by considering their Payne effect. On increasing the MA content, scorch and cure times increased along with maximum torque and torque difference. The MNR with 4 phr of MA exhibited the least filler–filler interactions, as indicated by the retention of the storage modulus after applying a large strain to the filled compound. This MNR compound also provided the highest tensile strength among the cases tested. It is interesting to highlight that MNR, with an appropriate MA content, reduces filler–filler interactions, and, thereby, enhances the HNT filler dispersion, as verified by SEM images, leading to improved mechanical and dynamical properties.

Keywords: natural rubber; maleated natural rubber; maleic anhydride; Halloysite Nanotubes

1. Introduction

Comparatively small loading levels of nanofillers in rubber matrices have drawn considerable attention during recent decades [1]. The physical properties of rubber can be clearly improved by this means. Such improvements depend on several factors, such as the filler aspect ratio, the degree of dispersion, and the filler orientation. Halloysite Nanotubes (HNT) are interesting nanofillers that have recently become available [2–4]. A HNT is a unique and versatile nanomaterial that is formed by the physical weathering of aluminosilicate minerals, and is composed of aluminum, silicon, hydrogen and oxygen. A HNT has two interlayer surfaces, one with aluminum hydroxide (Al-OH) groups located inside the tubes, and the other with siloxane (Si-O-Si) groups that cover the outer surfaces of the HNT.

Due to the unique surface chemistry of HNT, the compatibility of natural rubber (NR) with HNT is of interest, and several studies have sought to overcome their incompatibility. They have used modified rubbers as compatibilizers, [5] coupling agents [6], and adjusted the steps of preparation [7]. The application of a compatibilizer has been observed to affect the overall structure of a composite. As for the modified natural rubber, the improved compatibility of NR with HNT was obtained through certain functional groups, i.e., maleic anhydride (MA). Pasbakhsh et al. [8] used MA grafted onto ethylene propylene diene rubber (EPDM-g-MA) for compatibility with HNT. It has been demonstrated that the HNT filler agglomerates less, and has a finer dispersion in the presence of EPDM-g-MA. This has been attributed to reactions between hydroxyl groups on HNT surfaces and succinic anhydride available at the EPDM-g-MA. Similar observation was also found by Ismail et al. [9], who proposed interactions between hydroxyl groups in paper sludge and MA groups of maleated natural rubber

(MNR). From a structural point of view, the interactions of NR with HNT can be made possible by introducing this type of modified natural rubber.

The aim of this study is to find the same solution, as it has been shown to provide remarkable properties to the composites. Therefore, the use of MNR as the matrix for improved compatibility with HNT filler was proposed. Based on the chemical structure of MNR, it was expected to have improved compatibility between the succinic anhydride group of MNR and silanol and/or siloxane groups of HNT, and to have improved other related properties of the composites. To date, no prior report has been published on detailed investigations concerning the use of MNR as a matrix with HNT filler. In this study, curing characteristics, mechanical properties, and dynamic properties were investigated and discussed. This study will improve the scientific understanding of how MNR could influence the properties of rubber/HNT composites, and will provide useful information for preparing rubber products based on MNR/HNT composites.

2. Materials and Methods

2.1. Materials

The Standard Thai Rubber (STR) 5L as the main natural rubber (NR) grade used for preparing the composites. It was purchased from Chalong Latex Industry Co., Ltd., Thailand. The halloysite nanotubes (HNT) was supplied by Imerys Tableware Asia Limited, New Zealand. The main components in the HNT were SiO_2 (49.5 wt%), Al_2O_3 (35.5 wt%), Fe_2O_3 (0.29 wt%), and TiO2 (0.09 wt%), with traces of CaO, MgO, K_2O and Na_2O. Maleic anhydride was supplied by Sigma-Aldrich (Thailand) Co., Ltd., Bangkok, Thailand. The activators zinc oxide (ZnO) and stearic acid were purchased from Global Chemical Co., Ltd., Samut Prakan, Thailand, and Imperial Chemical Co., Ltd., Bangkok, Thailand, respectively. The accelerator, N-cyclohexyl-2-benzothiazole sulfenamide (CBS) was bought from Flexsys America L.P., West Virginia, USA, and sulfur was purchased from Siam Chemical Co., Ltd., Samut Prakan, Thailand.

2.2. Preparation of MNR

The grafting of MA onto NR was prepared by mixing NR with MA in an internal mixer (Brabender Plasticorder) at a temperature of 145 °C for approximately 10 min with a rotor speed of 60 rpm. The MA contents varied within 1–8 phr. The resulting MNR was then purified to confirm the grafting of MA onto NR. This was carried out by dissolving the rubber sample in toluene at room temperature for 24 h and then at 60 °C for 2 h; the soluble part was collected and precipitated in acetone. The sample was dried in a vacuum oven at 40 °C for 24 h. The purified MNR was finally characterized for a Fourier Transform Infrared Spectroscopy (FTIR) spectrum.

2.3. Preparation of Rubber Composites

Table 1 lists the main ingredients used for preparing the rubber composites, in which the main matrix used was separated accordingly. The total amounts of additives were mixed in a Brabender (Plastograph® EC Plus, Mixer W50EHT 3Z, Duisburg, Germany), and, just after the dumping, the compounds were passed through a two-roll mill to avoid excess heat. The compounds were then compressed into specific shapes using a hydraulic hot press, with the vulcanizing times obtained by a moving-die rheometer (MDR), which is described later.

Table 1. Compounding ingredients used in the composites.

Ingredient	Amount (phr)	
	Control	MNR
NR	100	-
MNR *	-	100
ZnO	5	5
Stearic acid	1	1
CBS	2	2
Sulfur	2	2
HNT	10	10

Remark: * MNR used was compounded separately according to the MA content (e.g., 1–8 phr).

2.4. Observation of Functionality Changes Using Fourier Transform Infrared Spectroscopy (ATR-FTIR)

The FTIR spectra of MNR were analyzed using a Bruker FTIR spectrometer (Tensor27) with a smart durable-single-bounce diamond in the ATR cell. Each spectrum was recorded in transmission mode with 32 scans per spectrum and 4 cm^{-1} resolution from 4000 to 550 cm^{-1}.

2.5. Measurement of Curing Characteristics

The curing characteristics of the composites were determined using an MDR (Rheoline, Mini MDR Lite) at 150 °C. Torque, scorch time (ts_2), and curing time (tc_{90}) were determined according to ASTM D5289.

2.6. Measurement of Mechanical Properties

The samples were cut into a dumbbell shape according to ASTM D412. The tensile tests were carried out with a universal tensile machine (Tinius Olsen, H10KS, Pennsylvania, USA) at a cross-head speed of 500 mm/min. This was to determine 100% modulus, 300% modulus, tensile strength, and elongation at break. Further, the tear strength of the composites was also tested using the same machine by following the ASTM D624 with a cross-head speed of 500 mm/min. The tear strength recorded was the average of five repeated tests for each compound.

2.7. Dynamic Properties

The dynamic properties of the NR/HNT and MNR/HNT composites were studied using a Rubber Process Analyzer model D-RPA 3000 (MonTech Werkstoffprüfmaschinen GmbH, Buchen, Germany). The composite sample was cured at 150 °C based on the curing time from Rheoline Mini MDR Lite (Prescott Instrument Ltd., Gloucestershire, England). Then, the sample was cooled down to 60 °C, and varying strains in the range from 0.5% to 100% were applied at a frequency of 10 Hz. The raw outputs storage modulus (G′) and damping characteristic (tan δ) were recorded, and the rubber–filler interactions in the composites were monitored through the Payne effect. The Payne effect can be quantified as follows:

$$\text{Payne effect} = G'_f - G'_i \tag{1}$$

where G'_f is G′ at 100% strain, and G'_i is G′ at 0.5% strain. A larger Payne effect indicates poorer rubber–filler interactions.

2.8. Scanning Electron Microscopy

The morphology of the rubber sample was screened by a scanning electron microscope (Quanta 400) to gain the detailed information on the dispersion of HNT filler in both NR and MNR. To produce electrostatic charge during scanning, fractured samples were coated with gold palladium prior to be scanned.

3. Results and Discussion

3.1. Functionalities of Maleated Natural Rubber

The FTIR spectra and the peak assignments for MNR at various MA contents are shown in Figures 1 and 2, and the peak assignments are also listed in Table 2. Considering the FTIR spectra of the purified MNR, an intense and broad characteristic band at wavenumber 1787 cm^{-1} and a weak absorption band at 1875 cm^{-1} confirmed the existence of succinic anhydride groups grafted onto the NR molecules. The observed bands were responses to the symmetric and asymmetric carbonyl (C=O) stretching vibrations of succinic anhydride rings. Moreover, there was an important peak captured at wavenumber of 1723 cm^{-1} due to the formation of carbonyl groups of opened ring structure succinic anhydride. The peaks seen in this study were quite similar to previous results seen in the literature [10,11].

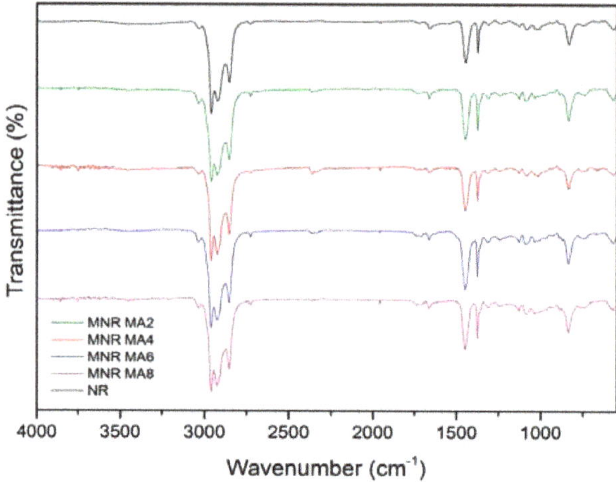

Figure 1. FTIR spectra of maleated natural rubber prepared at various maleic anhydride contents.

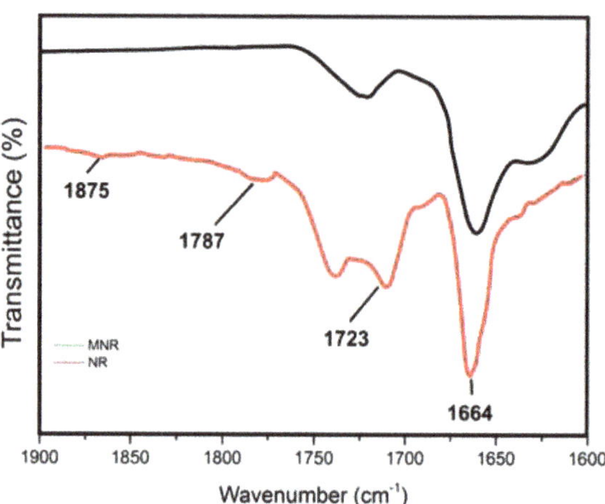

Figure 2. FTIR spectra in wavenumber range 1900–1600 cm^{-1} of NR and MNR at 8 phr of MA.

Table 2. The observed peaks and their assignments for MNR samples.

Wavenumber cm^{-1}	Assignment
2900	C-H stretch of NR
1875	C=O stretch of succinic anhydride (weak)
1787	C=O stretch of polymeric anhydride (weak)
1723	C=O stretch, carbonyl group
1664	C=C stretch of NR
835	C-H out of plane bend of NR

3.2. Curing Characteristics

The curing curves of the NR/HNT and MNR/HNT composites are shown in Figure 3, with the results summarized in Table 3. The minimum torque (M_L) slightly increased with the MA content in MNR, and M_L is known to represent the compound's viscosity. A further increase in MA concentration decreased the grafted MA content [12]. The MNR was melt-prepared, giving better contact with the rubber molecules than occurs in the solution state, as described by Nakason et al. [10]. A further increase of MA decreased the grafting efficiency, indicating the formation of maleate crosslinks under shear at an elevated temperature. This matches the higher M_L observed. A similar pattern was also found for the maximum torque (M_H) and for the torque difference ($M_H - M_L$) of the composites. This could be caused by differences in HNT dispersion and by different extents of crosslinking. Because these two torque values are known to represent the degree of crosslinking and/or the interactions within the composite system [13], this clearly indicates that a lower degree of crosslinking may be present due to acidity in the nature of maleic anhydride. As for the vulcanizing reaction, the MNR/HNT compounds display significantly longer scorch and cure times than the NR/HNT compounds. This is simply due the acidity in the nature coming from the ring opening reaction of succinic anhydride groups. Any chemical substance that makes the rubber compound more acidic will cause adsorption of accelerators [14] and retard the reactivity of accelerators.

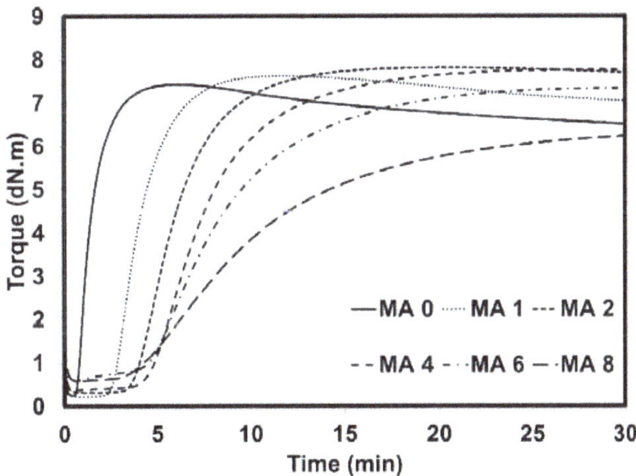

Figure 3. Tensile strength and elongation at break of NR/HNT and MNR/HNT composites.

Table 3. Curing characteristics of NR/HNT and MNR/HNT composites.

Compound	M_L (dN.m)	M_H (dN.m)	M_H-M_L (dN.m)	t_{S1} (min)	t_{c90} (min)	CRI (min^{-1})
MA 0	0.25	7.43	7.18	0.84	2.81	50.76
MA 1	0.22	7.63	7.41	2.92	6.40	28.74
MA 2	0.32	7.82	7.50	4.24	9.71	18.28
MA 4	0.38	7.77	7.39	5.14	13.05	12.64
MA 6	0.58	7.34	6.76	5.29	15.39	9.90
MA 8	0.58	6.25	5.67	5.46	19.16	7.30

3.3. Mechanical Properties

Figure 4 shows the stress–strain curves of the NR/HNT and MNR/HNT composites. The stress and strain values appear to differ between the NR and MNR as a matrix. From the stress–strain curves, it is possible to estimate the change point of the strain for each of the samples. Clearly, the strain at the onset of the stress upturn for the MNR/HNT composites is much lower than that of the NR/HNT composite, and the onset strain decreases with increasing MA content. This observation indicates that the use of MNR affects the stress–strain behavior of NR and lowers the strain at the onset of the stress upturn. Higher compatibility between MNR and HNT is responsible for these findings. Further, the area underneath the stress–strain curve was examined to confirm the compatibility of rubber and the filler. This indicates the toughness of a material [15]. The largest area underneath the curve corresponds to the greatest toughness. The MNR/HNT composites showed a greater area underneath the stress–strain curve than the NR/HNT composites, and, therefore, greater toughness.

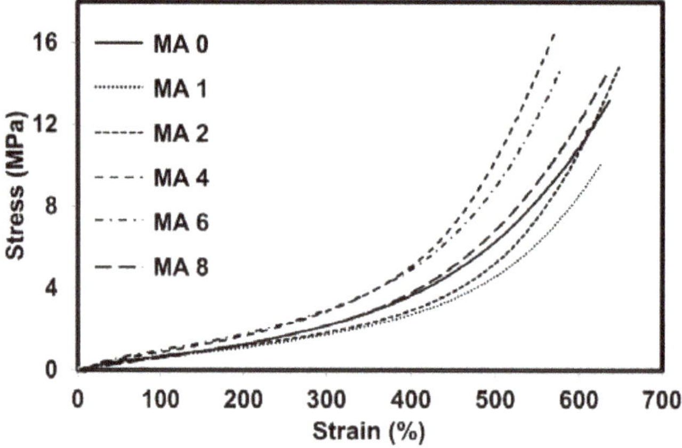

Figure 4. Stress–strain curves of the NR/HNT and MNR/HNT composites.

To focus in more detail, the tensile strength and elongation at break are plotted in Figure 5. The tensile strength increased when the concentration of MA increased. This is comparatively higher than for the NR/HNT composite. An increment in the tensile strength of the MNR/HNT composites is definitely caused by the improved compatibility of rubber and HNT. Grafting the succinic anhydride groups onto the NR molecules of the MNR enabled an increase in the polarity of the rubber and made it compatible with the HNT by forming the interactions between the succinic anhydride groups (either in the form of an opened ring or a cyclic structure) and the hydroxyl groups on the HNT surfaces. The formation of such interactions are seen and postulated in Figure 6. Similar interactions between the hydroxyl groups of HNT and the maleic anhydride groups have been discussed recently, such as in the case of EPDM-g-MA versus HNT by Pasbakhsh et al. [8].

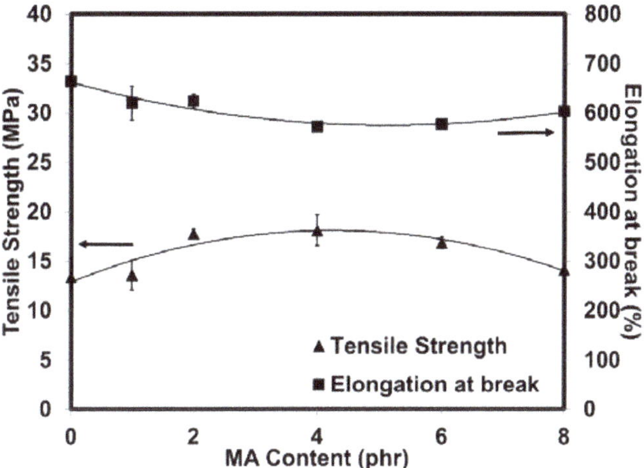

Figure 5. Tensile strength and elongation at break of NR/HNT and MNR/HNT composites.

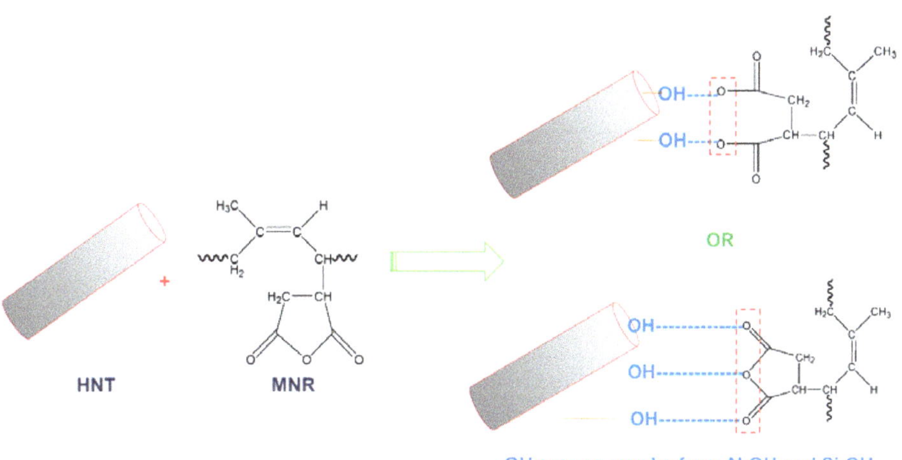

Figure 6. Possible interactions between MNR and HNT.

A further increase of the MA content also reduces the tensile strength since a maleate network may also be formed, and this would create a less uniform network structure [10]. Another possible reason may be the lower degree of crosslinking of the composites: a higher MA content can increase the acidity in nature of the MNR, leading to reduce the extension of crosslinking. This was clearly explained in M_H and $M_H - M_L$, reported in the preceding section. There was a similar effect observed in the elongation at break of the MNR/HNT composites. The strong interactions of NR and HNT can be confirmed from the stresses at 100% and 300% strains (see Figure 7). It can be seen that the stresses at 100% and 300% elongations (M100 and M300) increased with the MA content. As more MA was grafted onto the NR, more interactions took place, resulting in stiffer and harder composites. The reinforcement index (i.e., the ratio of modulus at 300% strain to modulus at 100% strain) is also embedded in Figure 7. Here, it can be seen that the reinforcing index gradually increased with the MA content. This is attributed to the higher compatibility between rubber and the filler when MNR was used as matrix. However, the lower tensile modulus and reinforcing index at a higher amount of MA may be simply due to the lower degree of crosslinking, as related to the torque difference found in the previous result. Furthermore, the tear strength showed a remarkable improvement with the MA content, as can be seen in Figure 8. The improved tear strength is due to the strong interactions of MNR and HNT as well as to the improved HNT dispersion in the rubber matrix.

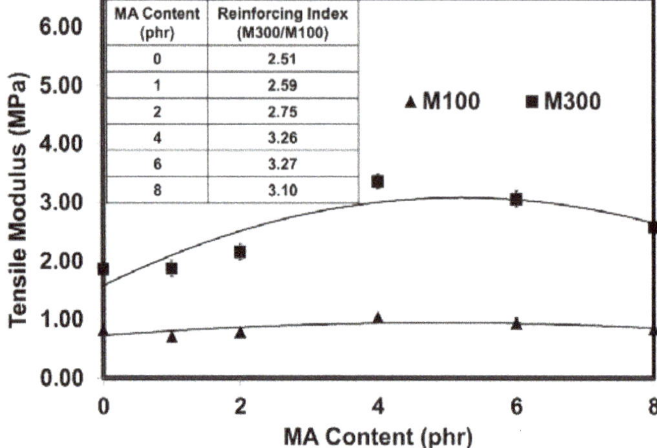

MA Content (phr)	Reinforcing Index (M300/M100)
0	2.51
1	2.59
2	2.75
4	3.26
6	3.27
8	3.10

Figure 7. Tensile modulus and reinforcing index of NR/HNT and MNR/HNT composites.

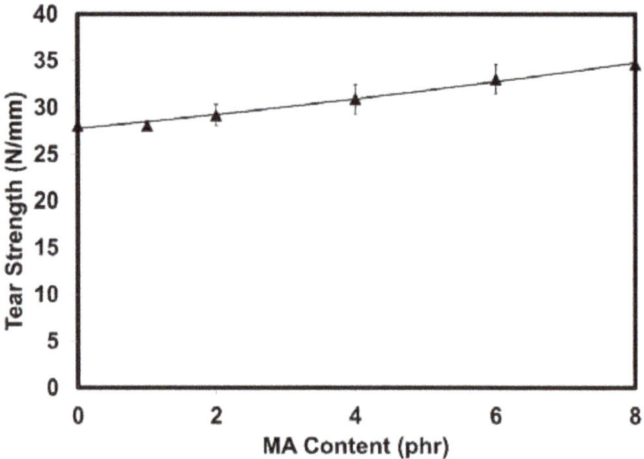

Figure 8. Tear strength of NR/HNT and MNR/HNT composites.

3.4. Dynamic Properties

The dynamic properties of the composites were determined using a Rubber Process Analyzer to investigate the storage modulus and the Payne effect. Figures 9 and 10 illustrate the storage modulus and the Payne effect of the NR/HNT and MNR/HNT composites. It can be seen that the storage modulus of all compounds was constant in the low-strain region, but decreased slightly with strains higher than 50%. This is common with viscoelastic materials and is due to the molecular stability of rubber. In addition to that, the Payne effect was estimated from the difference between the storage moduli at small and large strain amplitudes [16,17]. The level of Payne effect in the NR/HNT compound was 0.231 MPa and this decreased with the MA content to 0.174 MPa, 0.150 MPa and 0.160 MPa, with respect to the MPS loadings of 2–8 phr, in this order. This is a good indication that the interactions between MNR and HNT were improved. A lesser Payne effect indicates fewer filler–filler interactions [18]. However, higher Payne effect at higher amount of MA was linked to the lower degree of crosslinking observed previously. The damping characteristics (tan δ) as functions of strain are shown in Figure 11. It is obvious that the composites had lower damping characteristics with the addition of MA in the MNR, indicating a considerable degree of molecular mobility. This is simply due to the better interactions between rubber and the filler when using MNR as the main rubber matrix. The good compatibility with the HNT filler increases the adhesion at interface, resulting in an improvement in the elastic properties of the composites.

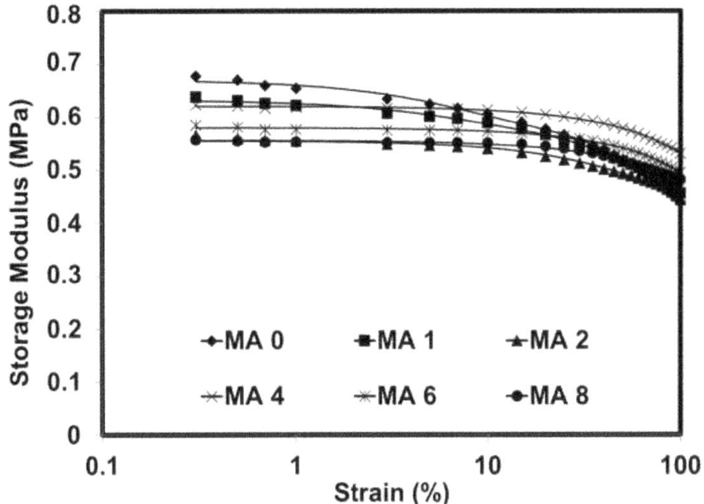

Figure 9. Storage moduli of NR/HNT and MNR/HNT composites.

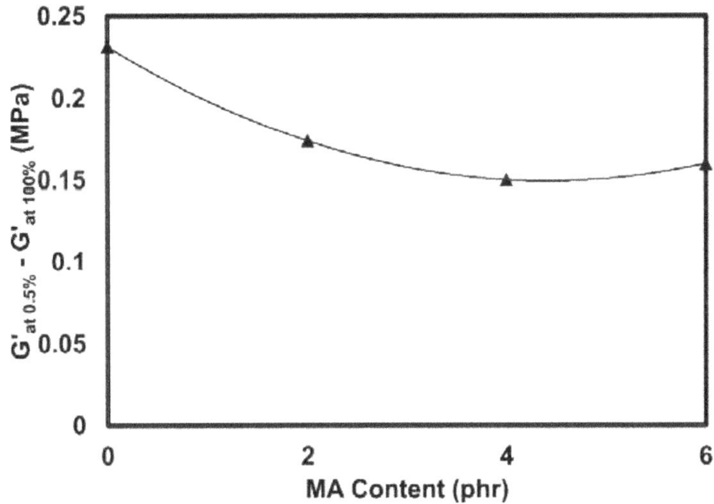

Figure 10. Payne effect of NR/HNT and MNR/HNT composites.

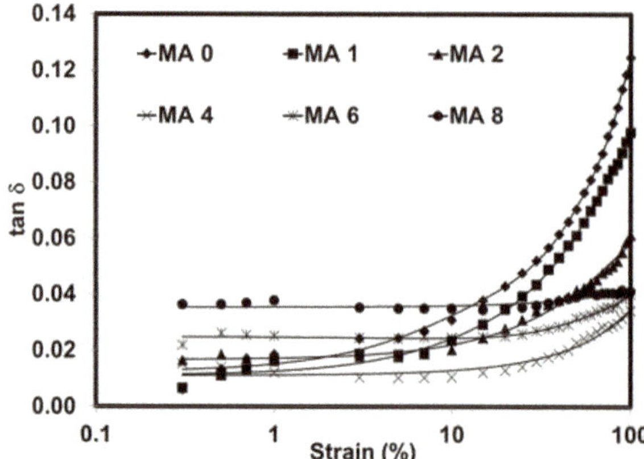

Figure 11. Damping characteristics (tan δ) of NR/HNT and MNR/HNT composites.

3.5. Scanning Electron Microscopy

Figures 12 and 13 show the tensile fractured surfaces of the NR/HNT and MNR/HNT composites. With low magnification in the images, the surface roughness of the composites was captured and is discussed subsequently. The scanning electron microscope (SEM) images of the composites with the NR and MNR as a matrix are shown in Figure 12A,B. Figure 12B shows a rougher surface and more matrix tearing lines than Figure 12A. This might be simply due to a higher rubber–filler interaction when the MNR was replaced in the formulation, increasing resistance to the propagation of cracks and thus providing a higher tensile strength in comparison to the NR/HNT composite. When a high MA content was grafted onto the NR (see Figure 12C,D), the roughness became more visible, indicating coherence of the rubber and filler, and, hence, the improved mechanical properties. In particular, Figure 12C shows that the highest tensile strength matched a uniformly homogenous pattern of roughness throughout the sample.

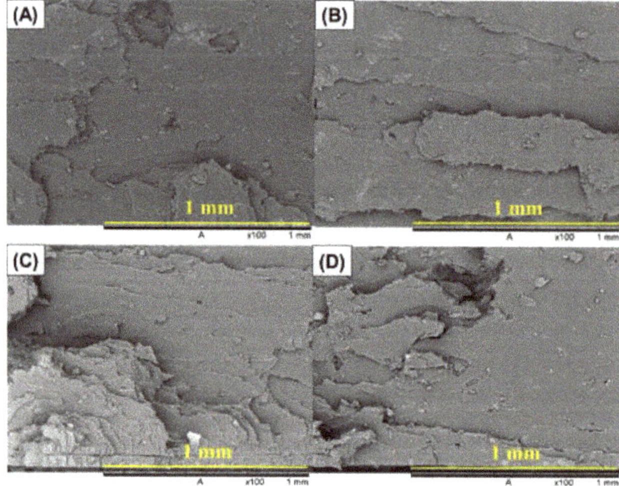

Figure 12. Scanning electron microscope (SEM) images of the tensile fracture surfaces of NR/HNT composites in the presence of MNR/HNT as dual compatibilizers: reference (**A**), MPS 0 phr (**B**), MPS 0.5 phr (**C**), and MPS 1.5 phr (**D**), all at 100× magnification.

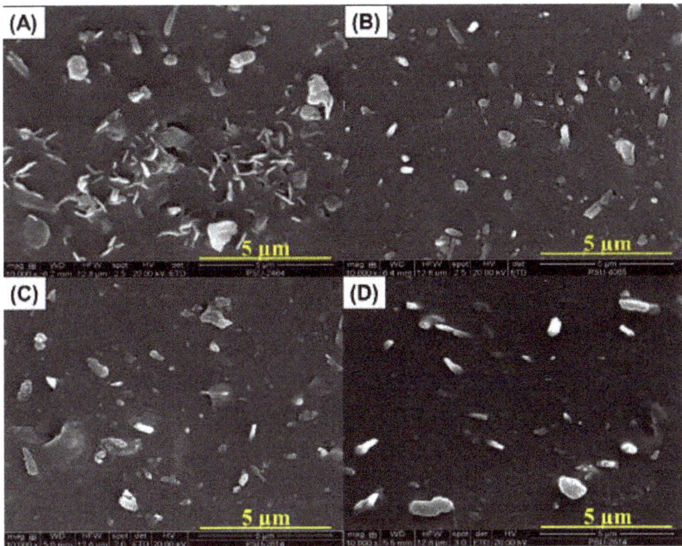

Figure 13. SEM images of the tensile fracture surfaces of NR/HNT composites in the presence of MNR/HNT as dual compatibilizers: reference (**A**), MPS 0 phr (**B**), MPS 0.5 phr (**C**), and MPS 1.5 phr (**D**), all at 10,000× magnification.

To assess the dispersion of HNT throughout the rubber matrix, SEM images were taken at 10,000× magnification and are shown in Figure 13. In Figure 13A, the agglomeration of HNT is seen on the fractured surface of the NR/HNT composite. When the NR matrix was replaced with MNR (Figure 13B), no agglomeration is seen due to good compatibility of the filler with the matrix. Moreover, the dispersion of HNT improved with the increased MA content. The homogeneity of the composites was significantly improved, especially when 4 phr and 8 phr MA was grafted onto NR (Figure 13C,D). The improved dispersion of HNT is clearly responsible for the improved tensile strength and tear strength, with more energy needed to break the samples. A better dispersion of HNT throughout the matrix increased the stress at a given strain. Similar observations on changes of microfractured surfaces with other fillers in NR composites in the presence of a compatibilizer have been reported previously [19].

4. Conclusions

The overall properties of the composites clearly improved when MNR was used as the main matrix of the HNT filled composites, when compared to those with plain NR. MNR has very special functional groups that can form hydrogen bonds with the hydroxyl groups available on the inner and outer surfaces of HNT. These interactions enhanced the tensile strength, modulus, and tear strength of the composites, and were corroborated by the decrease in Payne effect quantified from dynamic measurements. In this work, it has been clearly highlighted that the use of MNR improved HNT–rubber interactions and reduced filler–filler interactions, which benefits the mechanical and dynamical properties of the rubber composites. Therefore, MNR is a potentially alternative matrix for compatibility with the polar HNT surfaces, and could also improve the processing behavior without requiring the addition of a silane coupling agent, which is considered a complicated and costly method.

Author Contributions: Conceptualization, N.H.; methodology, N.H. and K.S.; software, Z.S.; validation, N.H., K.S. and H.I.; investigation, Z.S.; data curation, N.H.; writing—review and editing, N.H., K.S. and H.I.; supervision, N.H. All authors have read and agreed to the published version of the manuscript.

Funding: This research was funded by Prince of Songkla University through a Research Grant for New Scholar Grant No. SAT590650S and the APC was partially funded by Research and Development Office, Prince of Songkla University.

Acknowledgments: The authors would like thank Research and Development Office, Prince of Songkla University and Assoc. S.K. for assistance with manuscript preparation.

Conflicts of Interest: The authors declare no conflict of interest.

References

1. Arrighi, V.; McEwen, I.; Qian, H.; Prieto, M.S. The glass transition and interfacial layer in styrene-butadiene rubber containing silica nanofiller. *Polymer* **2003**, *44*, 6259–6266. [CrossRef]
2. Ismail, H.; Pasbakhsh, P.; Fauzi, M.A.; Bakar, A.A. Morphological, thermal and tensile properties of halloysite nanotubes filled ethylene propylene diene monomer (EPDM) nanocomposites. *Polym. Test.* **2008**, *27*, 841–850. [CrossRef]
3. Jia, Z.; Luo, Y.; Guo, B.; Yang, B.; Du, M.; Jia, D. Reinforcing and flame-retardant effects of halloysite nanotubes on LLDPE. *Polym. Plas. Technol. Eng.* **2009**, *48*, 607–613. [CrossRef]
4. Vahedi, V.; Pasbakhsh, P.; Chai, S.-P. Toward high performance epoxy/halloysite nanocomposites: New insights based on rheological, curing, and impact properties. *Mater. Des.* **2015**, *68*, 42–53. [CrossRef]
5. Paran, S.; Naderi, G.; Ghoreishy, M. XNBR-grafted halloysite nanotube core-shell as a potential compatibilizer for immiscible polymer systems. *Appl. Surf. Sci.* **2016**, *382*, 63–72. [CrossRef]
6. Rooj, S.; Das, A.; Thakur, V.; Mahaling, R.; Bhowmick, A.K.; Heinrich, G. Preparation and properties of natural nanocomposites based on natural rubber and naturally occurring halloysite nanotubes. *Mater. Des.* **2010**, *31*, 2151–2156. [CrossRef]
7. Du, M.; Guo, B.; Lei, Y.; Liu, M.; Jia, D. Carboxylated butadiene–styrene rubber/halloysite nanotube nanocomposites: Interfacial interaction and performance. *Polymer* **2008**, *49*, 4871–4876. [CrossRef]
8. Pasbakhsh, P.; Ismail, H.; Fauzi, M.A.; Bakar, A.A. Influence of maleic anhydride grafted ethylene propylene diene monomer (MAH-g-EPDM) on the properties of EPDM nanocomposites reinforced by halloysite nanotubes. *Polym. Test.* **2009**, *28*, 548–559. [CrossRef]
9. Ismail, H.; Rusli, A.; Rashid, A.A. Maleated natural rubber as a coupling agent for paper sludge filled natural rubber composites. *Polym. Test.* **2005**, *24*, 856–862. [CrossRef]
10. Nakason, C.; Kaesaman, A.; Supasanthitikul, P. The grafting of maleic anhydride onto natural rubber. *Polym. Test.* **2004**, *23*, 35–41. [CrossRef]
11. Sahakaro, K.; Beraheng, S. Reinforcement of maleated natural rubber by precipitated silica. *J. Appl. Polym. Sci.* **2008**, *109*, 3839–3848. [CrossRef]
12. Nabil, H.; Ismail, H.; Azura, A. Recycled polyethylene terephthalate filled natural rubber compounds: Effects of filler loading and types of matrix. *J. Elast. Plas.* **2011**, *43*, 429–449. [CrossRef]
13. Rattanasom, N.; Saowapark, T.; Deeprasertkul, C. Reinforcement of natural rubber with silica/carbon black hybrid filler. *Polym. Test.* **2007**, *26*, 369–377. [CrossRef]
14. Coran, A. Chemistry of the vulcanization and protection of elastomers: A review of the achievements. *J. Appl. Polym. Sci.* **2003**, *87*, 24–30. [CrossRef]
15. Nun-anan, P.; Wisunthorn, S.; Pichaiyut, S.; Nathaworn, C.D.; Nakason, C. Influence of nonrubber components on properties of unvulcanized natural rubber. *Polym. Adv. Technol.* **2020**, *31*, 44–59. [CrossRef]
16. Payne, A.; Whittaker, R. Low strain dynamic properties of filled rubbers. *Rubb. Chem. Technol.* **1971**, *44*, 440–478. [CrossRef]
17. Kaewsakul, W.; Sahakaro, K.; Dierkes, W.K.; Noordermeer, J.W. Cooperative effects of epoxide functional groups on natural rubber and silane coupling agents on reinforcing efficiency of silica. *Rubb. Chem. Technol.* **2014**, *87*, 291–310. [CrossRef]

18. Rooj, S.; Das, A.; Stöckelhuber, K.W.; Wang, D.-Y.; Galiatsatos, V.; Heinrich, G. Understanding the reinforcing behavior of expanded clay particles in natural rubber compounds. *Soft. Mat.* **2013**, *9*, 3798–3808. [CrossRef]
19. Waesateh, K.; Saiwari, S.; Ismail, H.; Othman, N.; Soontaranon, S.; Hayeemasae, N. Features of crystallization behavior of natural rubber/halloysite nanotubes composites using synchrotron wide-angle X-ray scattering. *Intern. J. Polym. Anal. Charac.* **2018**, *23*, 260–270. [CrossRef]

Article

Wastewater Treatment from Lead and Strontium by Potassium Polytitanates: Kinetic Analysis and Adsorption Mechanism

Anna Ermolenko [1], Alexey Shevelev [1], Maria Vikulova [1], Tatyana Blagova [1], Sergey Altukhov [1], Alexander Gorokhovsky [1], Anna Godymchuk [2,3], Igor Burmistrov [1,4,5,*] and Peter Ogbuna Offor [6]

1. Department of Chemistry and Technology of Materials, Yuri Gagarin State Technical University of Saratov, 77 Polytecnicheskaya Street, 410054 Saratov, Russia; molish01@mail.ru (A.E.); titans5@rambler.ru (A.S.); vikulovama@yandex.ru (M.V.); blagova-ta@yandex.ru (T.B.); sergeypavlovicha@yandex.ru (S.A.); algo54@mail.ru (A.G.)
2. Department of Materials Science, National Research Tomsk Polytechnic University, 30 Lenina Avenue, 634050 Tomsk, Russia; godymchuk@mail.ru
3. Tobolsk Complex Scientific Station, Ural Branch of the Russian Academy of Science, 15 Osipov Academician Street, 626152 Tobolsk, Russia
4. Department of Functional Nanosystems and High-Temperature Materials, National University of Science and Technology "MISiS", 4 Leninsky Avenue, 119049 Moscow, Russia
5. Engineering Center, Plekhanov Russian University of Economics, 36 Stremyanny Lane, 117997 Moscow, Russia
6. Metallurgical and Materials Engineering Department, University of Nigeria, Nsukka 410001, Nigeria; peter.offor@unn.edu.ng

* Correspondence: glas100@yandex.ru; Tel.: +7-917-201-87-03

Received: 23 December 2019; Accepted: 7 February 2020; Published: 12 February 2020

Abstract: The reduction of heavy and radioactive metal pollution of industrial wastewater remains a vital challenge. Due to layered structure and developed surface, potassium polytitanate had potential in becoming an effective sorbent for metal extraction from wastewater in the presented paper. On the basis of the different sorption models, this paper studied the mechanism of Pb^{2+} and Sr^{2+} cation extraction from aqueous solution by non-crystalline potassium polytitanate produced by molten salt synthesis. The ion exchange during metal extraction from model solutions was proven by kinetic analysis of ion concentration change, electronic microscopy, and X-ray fluorescence analysis of sorbent before and after sorption, as well as by theoretical modeling of potassium, lead, and strontium polytitanates. The sorption was limited by the inner diffusion in the potassium polytitanate (PPT) interlayer space, as was shown using the Boyd diffusion model. The sorption processes can be described by Ho and McKay's pseudo-second-order model compared to the Lagergren pseudo-first-order model according to kinetic analysis. It was found that the ultimate sorption capacity of synthesized sorbent reached about 714.3 and 344.8 (ions mg/sorbent grams) for Pb^{2+} and Sr^{2+} ions, respectively, which was up to four times higher than sorption capacity of the well-known analogues. Therefore, the presented study showed that potassium polytitanate can be considered as a promising product for industry-scaled wastewater purification in practice.

Keywords: sorption; heavy metals; radioactive metals; potassium titanate; ion exchange

1. Introduction

Industrialization has caused the pollution of natural waters by industrial wastes containing heavy and radioactive metals from ore-beneficiation plants [1], metal treatment [2], paints and pigments [3], and batteries [4]. Lead and radioactive strontium are toxic and cause environmental and human health damage [5]. Accumulation of heavy metals in the reservoirs closed by enterprises may bring severe hazards to the bio-system and society [6]; therefore, the reduction of the amount of metal in the wastewater remains a vital challenge.

Ion-exchange sorption is still one of the most effective and cost-beneficial methods for removing heavy metals from aqueous media [7–9], as it provides a high purification degree (up to 100%) of highly polluted waters, the possibility of sorbent regeneration, and simple use. The most common heavy metal adsorbents are synthetic zeolite, natural kaolinite, chitin, chitosan [10–14], carbon nano-tubes (CNT), carbon materials [15–17], agricultural waste such as rice bran, orange peel [18], montmorillonite, industrial by-products (such as lignin, diatomite, clino-pyrrhotite, lignite, aragonite shells, peat), bio-sorbents [19], and nanomaterials of metal oxides [20]. However, the introduction of most materials into the practice is limited because of its' high cost, sensitivity to occupational conditions, and complicated re-use. Therefore, the search for effective sorption materials is still important.

Potassium titanates have been considered as promising sorbents for heavy metals. In particular, sorption capacity of crystalline potassium hexatitanate ($K_2Ti_6O_{13}$) was found to reach 0.80 mmol/g in Cd^{2+} solution [21]; meanwhile, the capacity of crystalline potassium tetratitanate ($K_2Ti_4O_9$) in Cu^{2+} solution was 1.94 mmol/g [22]. The crystalline sodium titanate proved the capacity of 2.66 mmol/g compared to the 0.24 mmol/g for $H_2Ti_4O_9$ and 0.16 mmol/g for $H_2Ti_3O_7$ [22–24]. Besides crystalline potassium titanates, the special interest of researchers has been taken to potassium polytitanates (PPT). Due to the high surface area and the relatively large distance between the layers of the titanium–oxygen octahedron (1.5–2.0 nm), PPTs have increased ion-exchange capacity for heavy metal removal from wastewater [25,26]. Moreover, PPTs have a negatively charged surface, which can electro-statically attract metal oxide-hydroxide cations, as shown in our previous investigations of layered lepidocrocite-like quasi-amorphous compounds based on modified PPTs extracting organic dyes [27] and nickel ions [28].

Despite numerous studies on quantitative assessment of crystalline potassium titanate sorption properties, there is still a lack of knowledge about the sorption mechanism and ultimate PPT sorption capacity. Thus, the finding of this current study provides kinetic analysis to establish the mechanism of heavy metal ion sorption on a new and poorly studied type of X-ray amorphous sorbents—potassium polytitanates.

2. Materials and Methods

2.1. Synthesis and Characterization of PPT Sorbent

Potassium polytitanate (PPT) was synthesized by molten salt synthesis [25]. For the synthesis, the following chemicals were used for the mixture preparation: KOH (98% purity, Vekton, St. Petersburg, Russia), KNO_3 (98% purity, Reahim, Moscow, Russia), and TiO_2 (99% anatase, Aldrich, Darmstadt, Germany, CAS 13463-67-7). The dry mixture was stirred for 5 min in an alundum crucible with components mass ratio $KOH/KNO_3/TiO_2$ = 30:40:30 at 25 ± 2 °C. Then, after adding distilled water in the mass of twice the TiO_2 mass, it was stirred again for 5 min. Further, the mixture was heated with a heating rate of 7 °C/min up to 500 ± 10 °C and was held for 3 h in an electric muffle furnace PM-8 (TD Lab-Therm Ltd., Moscow, Russia) prior to natural cooling in the closed furnace for 24 h. The cooled product was triturated in an agate mortar and placed in a glass beaker filled with distilled water (powder/water ratio = 1:2) for thorough mixing. The resulting suspension was washed six times with distilled water by decanting until the pH of the washing water was 10.5 ± 5. The pH of the suspension was monitored with a 150MI pH meter (Measuring technology Ltd., Moscow, Russia). Then, the suspension was filtered and dried at 50 ± 0.5 °C in a Binder FD 53 drying oven (BINDER GmbH, Tuttlingen, Germany) for 24 h. The dry sample was ground into a powdered material in an agate mortar. Considering the material's hydrophilicity and the possibility of humidity change during the long-term storage, all measurements were carried out for a freshly synthesized powder (no more than 1 week after synthesis).

The morphology and elemental analysis of the sorbent surface were performed with the use of a scanning electron microscope (SEM) with integrated energy-dispersive X-ray spectroscopy (EDX) analyzer EXplorer (ASPEX, Delmont, USA). After spraying gold onto a sorbent placed on an aluminum substrate, measurements were carried out at the accelerating voltage of 15 kV. The individual particles of the synthesized titanates were studied in terms of morphology and size by a transmission electron microscope (TEM) JEOL JEM 1400 with an accelerating voltage of 120 kV (JEOL Ltd., Tokyo, Japan). The PPT-specific surface was determined with the Quantachrome Nova 2200 analyzer (Quantachrome Instruments, Boynton Beach, USA) by using the initial part of the isotherm of physical sorption of high pure nitrogen (99.999%) and was calculated by instrumental software using the Brunauer-Emmett-Teller (BET) model. Preliminary degassing of the samples was carried out at 150 ± 2 °C for 3 h. Own instrumental error did not exceed 2%.

The phase composition of the synthesized sorbent was determined using an X-ray diffractometer ARL X'TRA (Thermo Scientific, Ecublens, Switzerland) with CuKα radiation (λCuKα = 0.15412 nm) in a 2Θ angle range from 5 to 60 degrees. Bragg–Brentano measurement geometry was used for analysis with step-by-step scanning mode, speed of 2 degrees, and signal accumulation time of 1 s. The library of the international electronic database of diffraction standards (produced by ICDD—International Center for Diffraction Data)—PDF-2 (Powder Diffraction File-2) database in the Crystallographic Search-Match Version 3,1,0,2 B was used for phase identification on the resulting diffractograms. The permissible absolute error limits in measuring the angular positions of the diffraction maximum were ±0.0015°. Particle-size distributions of the ceramic filler and polymer particles in the dispersion were obtained using Analysette 22 MicroTec Plus Fritsch laser diffraction equipment.

The performed quantum modeling of the PPT structure facilitated the understanding of the material structure and the explanation of possible sorption mechanisms. Thus, optimizing to minimum potential energy, the clusters $K_4Ti_8O_{18}$, $Pb_2Ti_8O_{18}$, and $Sr_2Ti_8O_{18}$ were modeled. As a modeling result, the calculated "cation-oxygen" distances in the PPT structure were obtained.

A quantum chemical study was performed by the Priroda 6 program, using the density functional theory (DFT), Perdew–Burke–Ernzerhof (PBE) functional [29], L1 basis, using scalar relativistic corrections [30,31] prior to the calculation of the cation-oxygen inter-atomic distances in the compounds. The obtained values allowed for the creation of the PPT cluster by incrementally changing the geometry of the molecules with optimization of the system potential energy to the minimum.

2.2. Sorption Capacity Assessment

Lead nitrate ($Pb(NO_3)_2$, 99.5% pure, Vekton Ltd., Russia) and strontium 6-water chloride ($SrCl_2{\cdot}6H_2O$, CAS 10025-70-4 Vekton Ltd., Russia) were used to prepare two separate aqueous solutions (for a separate measurement of lead and strontium ions sorption) with Pb^{2+} or Sr^{2+} concentration of 50 mg/mL (hereinafter in the calculations—C_0). The solutions were prepared on the basis of distilled water (pH = 6.0–6.5) right before the experiment. The pH of the initial solutions were 4.50 ± 0.20 and 10.65 ± 0.30, respectively, for Pb^{2+} and Sr^{2+} solutions. The PPT was exposed to the prepared solution (sorbent concentration 110 g/L) at 25 ± 0.2 °C for 90 min with constant stirring using a magnetic stirrer (60 rpm). Throughout the experiment, the pH of the system was measured using an MA130 ion-meter (Mettler-Toledo GmbH, Greifensee, Switzerland).

Every 5 min, 5 mL of the suspension was taken and filtered on a Bola GmbH PTFE filter (Bohlender GmbH, Grünsfeld, Germany, pore size—10 μm, pressure—50 mm Hg). The concentration of metal ions (ion concentration at the given point in time, C_t) was determined in the filtrate using the X-ray fluorescence method on the Spectroscan-MAX-G spectrometer (Spectron, Moscow, Russia) with a scanning crystal diffraction channel. The measurements were carried out at 20 °C using an X-ray tube with a silver anode (4 W, 40 kV, and 100 μA) and LiF crystal analyzer (200) in the interval—810...3200 mA with a scanning step of 2.0 mA. The method sensitivity for metal ions was 1–20 ppm. Kinetic curves (($\Delta C = f(t)$, where $\Delta C = C_0 - C_t$ is the change in concentration by the time t, min) were created on the basis of obtained data. Reaching the constant ΔC value means the occurrence of the dynamic equilibrium corresponding to the metal ions concentration C_1. The sorption capacity Q (mg of metal/g of sorbent) was calculated by Equation (1):

$$Q = \frac{(C_0 - C_t)V}{m} \tag{1}$$

where C_0 is the initial metal ion concentration, g/L; C_t is the metal ion concentration after the sorbent/sorbate dynamic equilibrium onset, g/L; V is the volume of the metal salt solution, l; and m is the mass of sorbent, g.

2.3. Numerical Analysis of Adsorption Kinetics

Sorption kinetics analysis was performed by modeling after a detailed comparison of several key models to evaluate the contribution of various sorption mechanisms. Boyd's diffusion model is currently the only kinetic model allowing simultaneous examination of external and internal diffusion as limiting sorption stages. This approach simplifies the analysis by eliminating the need for multiple recalculations of experimental data using several models.

By the Boyd model, external diffusion processes were characterized by the time dependence $-\log(1-F)$, where F is the equilibrium degree in the system calculated by Equation (2):

$$F = \frac{Q_t}{Q_e}, \tag{2}$$

where Q is the adsorbed substance amount per sorbent mass unit at the moment of time t (Q_t) and in the equilibrium state (Q_e, mg/g).

The sorption limitation by intra-diffusion processes was determined by the $F - Bt$ dependence, where Bt is the dimensionless Boyd parameter calculated more accurately by Equation (3):

$$Bt = 0.4977 - \ln(1 - F) \tag{3}$$

The identification of the influence of the chemical stage of ion exchange on the process of PPT interaction with metal ions was performed considering sorption using the kinetic model of pseudo-first and pseudo-second Lagergren's orders, which takes into account the solid sorbent sorption capacity [32], described by the Equation (4):

$$\log(Q_e - Q_t) = logQ_e - \frac{kt}{2.303} \tag{4}$$

where k is the sorption rate constant, min^{-1}.

The correspondence of the obtained experimental data of the pseudo-second-order model was performed using the Ho and McKay classical equation [33] in the linear form:

$$\frac{t}{Q_t} = \frac{1}{k_2 Q_e^2} + \frac{t}{Q_e} \tag{5}$$

where k_2 is the rate constant, $g{\cdot}mmol^{-1}{\cdot}min^{-1}$.

3. Results and Discussion

3.1. Characterization of Sorbents

The morphology, structure, and elemental and phase composition of PPT was studied before and after sorption. The synthesized PPT was a powder with particle size distribution from 100 nm to 10 μm (Figure 1a,b). Large particles were mainly aggregates of individual particles, having a layered structure and consisting of flat flakes with sizes mainly from 300 to 1000 nm and thickness of several atomic layers (Figure 1c). The specific surface area of the obtained PPT was from 90 to 100 m^2/g (variation in the series of experiments). X-ray diffractogram of the PPT before sorption (Figure 2b) did not include specific peaks, whereas the wide reflex at 48 degrees indicated the formation of the PPT structure [25,34]. The broad peak, amorphous halo, in the range between 25 and 35 degrees of 2Θ testified the X-ray amorphous structure of PPTs and the absence of long-range order in the lattice periodicity.

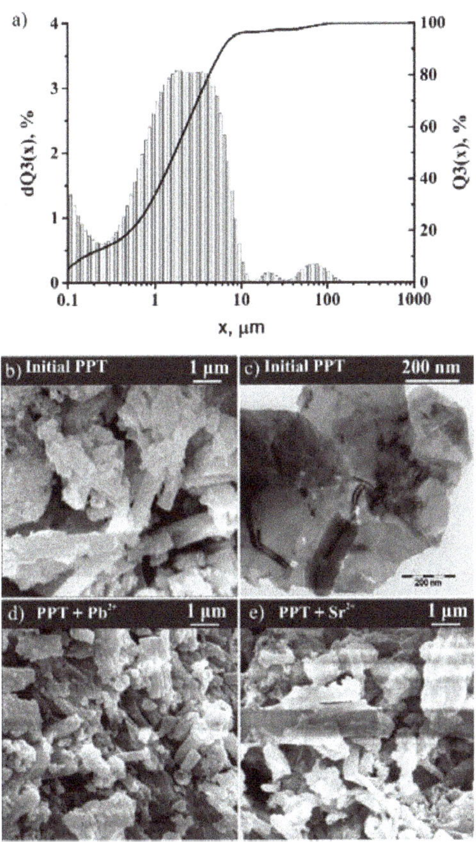

Figure 1. Particle size distribution of potassium polytitanates (PPT) before sorption (**a**); scanning electron microscope (SEM) (**b**) and transmission electron microscope (TEM) image (**c**) of PPT before sorption, and SEM image of PPT after sorption of Pb^{2+} (**d**) and Sr^{3+} (**e**).

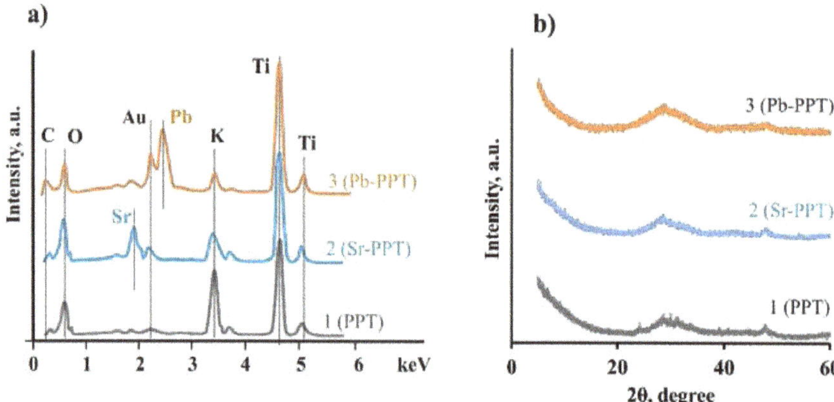

Figure 2. Chemical composition data (energy-dispersive X-ray spectroscopy (EDX)) (**a**) and X-ray (**b**) of synthesized PPT and exposed sorbents (Sr-PPT and Pb-PPT).

It was shown that the particle morphology of exposed sorbents did not differ from the initial form (Figure 1d,e). The adsorption intensity of the K band decreased, whereas Pb and Sr bands appeared according to its X-ray fluorescence spectrum (Figure 2a). The obtained data suggest the ion exchange of K^+ ions with Pb^{2+} and Sr^{2+} on the PPT surface. In the series of K^+, Pb^{2+}, and Sr^{2+} ions, the bond length between cat-ion and oxygen, calculated using the density functional theory, decreased in the structure of the PPT (Table 1). The obtained values of the bond length were relatively close to those found in the literature data [35]. The discrepancy with the literature data was due to the limited size of the calculated cluster and estimated errors of the optimization of the system potential energy to the minimum.

Table 1. Inter-atomic distance "cation-oxygen" for $K_4Ti_8O_{18}$ and $Me_2Ti_8O_{18}$ (Me = Pb or Sr).

Complex	Cation	Calculated Distance "Cation-Oxygen", nm	Reviewed Distance "Cation-Oxygen" [35], nm	Divergence with Literary Values, %
$K_4Ti_8O_{18}$	K^+	0.280	0.269	4
$Pb_2Ti_8O_{18}$	Pb^{2+}	0.232	0.262	11
$Sr_2Ti_8O_{18}$	Sr^{2+}	0.261	0.256	2

On the basis of the smaller inter-atomic distance indicating a stronger bond of Pb^{2+} and Sr^{2+} ions with oxygen compared to K^+, it can be assumed that the strong chemical connection of Pb^{2+} and Sr^{2+} ions would minimize the probability of desorption of ions and, therefore, prevent secondary environmental pollution.

3.2. Change of Sorption Rate

Figure 3 shows the time- and pH-dependent kinetics and limited sorption capacity of Pb^{2+} and Sr^{2+} ions. According to the kinetic curve of Pb^{2+}sorption, the induction period accompanied the sorbent's aging during the first 5 min of being on (Figure 3a, Pb^{2+}-ion curve). This effect may have been caused by the slow destruction of inert lead complexes followed by its accumulation near the surface, which is typical of transition metal solutions [36,37]. Then, the lead sorption in the solution started to intensify, and in 30 min the system achieved a quasi-equilibrium state at $\Delta C = 42.0 \pm 1.9$ mg/mL. The second equilibrium occurred at 45 min, when the efficiency of Pb^{2+} ion extraction was 99.4% (the maximum decrease in concentration was 49.7 ± 1.5 mg/mL). The complex mechanism of PPT interaction with Pb^{2+} ions was caused by the layered structure of the sorbent—the first equilibrium at the initial stage could be most likely be explained by the ion adsorption on the PPT surface, whereas the second equilibrium was related to sorbate diffusion into the PPT interlayer space. Because the pH of the Pb^{2+}-based PPT suspension during the sorption process was within the range of 4.5–6.2 (Figure 3b), sedimentation of oxide–hydroxide metal complexes was not expected, and Pb^{2+} ion intercalation processes were very possible.

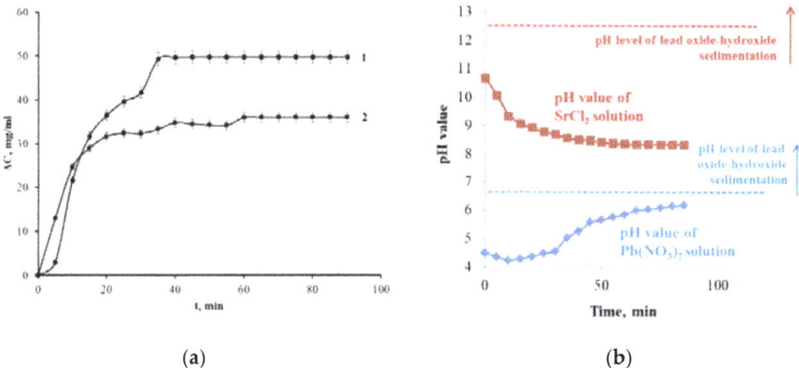

Figure 3. Sorption kinetic curve of Pb^{2+} (1) and Sr^{2+} (2) ions on PPT at pH < 6.5 and pH > 8.5, respectively (**a**), and the pH value during the Pb^{2+} and Sr^{2+} sorption on PPT (**b**).

When extracting Sr^{2+} ions in the first 10 min, the concentration of the ion changed linearly to $\Delta C = 25.0 \pm 1.6$ mg/mL (Figure 3, Sr^{2+} ion curve). Then, it increased gradually, and after 45 min reached a maximum value of 36.0 ± 1.0 mg/mL; at that point, the degree of extraction reached 72%. This kinetic curve behavior was likely due to the twice larger ionic radius of Sr^{2+} than Pb^{2+}, complicating the diffusion of the ion into the PPT interlayer space. In this regard, there was a single equilibrium state related to the ion sorption on the sorbent surface. Moreover, the pH of the Sr^{2+}-based PPT suspension during the sorption process exceeded 8.5 (Figure 3b), meaning that the basic medium contributed to the formation of surface oxide–hydroxide metal complexes, and excluded the intercalation process.

However, the kinetic analysis did not explain the differences in sorption of cations, and it did not reveal the mechanism of sorption combining the transfer of cat-ions and chemical interaction of PPT with metal ions.

3.3. *Analysis of the Diffusion Component of the Sorption Process (Boyd Model)*

To assess the role of cations transfer to and inside the solid surface, we used the diffusion model (Boyd model). Within the Boyd diffusion models, a quantitative approach was applied for the primary distinction between intra- and external-diffusion adsorption limitation, which involved the analysis of kinetic data in the coordinates $-\ln(1-F)$ and t.

According to the Boyd model, the linear dependence in the coordinates $(-\ln(1-F))$-t indicates the external diffusion of ions to and along the surface of the sorbent as the limiting stage of sorption. The coefficient of approximation accuracy for given dependences was relatively low over the entire time range for studied systems: $R^2 \approx 0.9$ for Pb^{2+} and $R^2 < 0.9$ for Sr^{2+} ions (Figure 4). However, the identification of the linear areas for both metal ions became possible within the first 25 min of interaction. Therefore, the Pb^{2+} sorption was limited by external (film) diffusion only within first contact stages (sorption time was less than 25 min). Then, while the PPT active centers were being filled, the impact of intra-diffusion mass transfer on the sorption process increased. The logarithmic dependence of the equilibrium attainment degree on the Boyd parameter (Figure 5) obtained for the sorption of Pb^{2+} ions proved the intra-diffusion mechanism of sorption. The high approximation confidence coefficient of the given curve of the logarithmic trend line ($R^2 > 0.98$) indicated a large contribution of the internal diffusion of metal ions in the sorbent interlayer space to the overall rate of the sorption process.

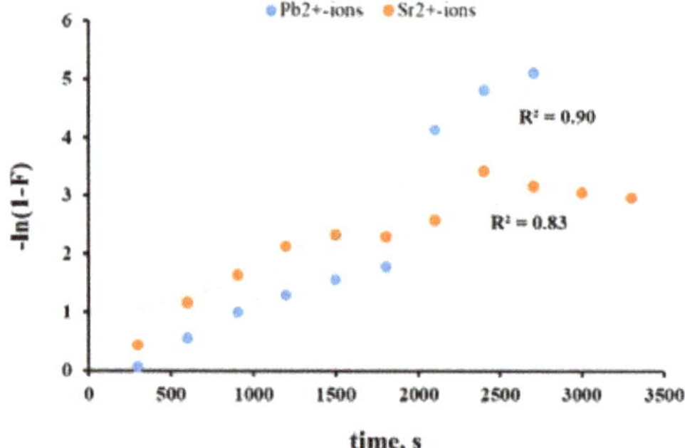

Figure 4. Change of $-\ln(1-F)$ coefficient for metal ion sorption on PPT.

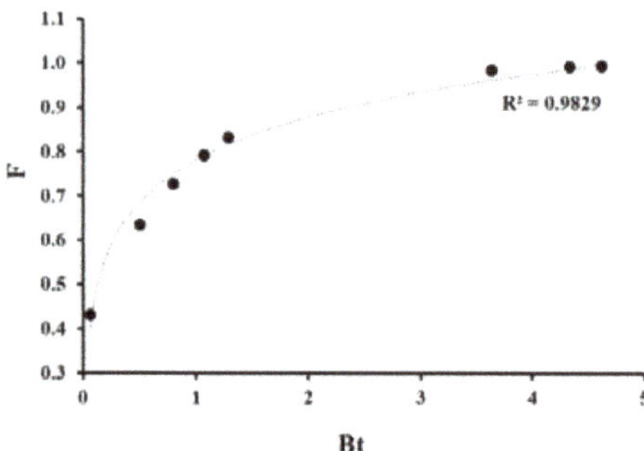

Figure 5. Change of equilibrium attainment degree (F) depending on Bt during Pb^{2+} ion sorption.

Analysis of the data from Figures 3–5 suggested a mixed kinetic mechanism (including components of external and internal diffusion) for the Pb^{2+} ion sorption at PPT, where the sorption process was limited by the stage of external diffusion of ions to and along the sorbent surface at the beginning of the interaction between metal ions and PPT, and then internal diffusion in the PPT interlayers became the slowest stage after reaching the certain saturation degree of sorption centers. Thus, the kinetic curve had two equilibrium parts related to adsorption and intercalation processes. However, the Sr^{2+} ion sorption rate at the first stage of interaction with PPT was limited only by the external diffusion of ions in the solution and on the surface of the sorbent (Figure 4) due to discrepancy of $F - Bt$ dependencies for the Sr^{2+} solution, proving insignificant impact of internal diffusion on the final sorption rate.

The use of the Boyd model showed a complex interaction of PPTs with Pb^{2+} ions, and the limiting stage of external and internal diffusion was difficult discover because the total sorption rate was determined by a different mechanism of diffusion transfer of metal cations.

3.4. *Analysis of Chemical Kinetics of Sorption Processes*

The kinetic models of pseudo-first and pseudo-second orders can allow for the estimation of the contribution of chemical interaction of sorbent and cation into the sorption kinetics. When analyzing the sorption process using a pseudo-first-order model (Figure 6), the linear dependencies of $\log(Q_e - Q_t)$ with a sufficiently low coefficient of R^2 were obtained. Furthermore, the R^2 for Sr^{2+} ions (Figure 6, Sr^{2+} ion curve) lay below R^2 for Pb^{2+} ions (Figure 6, Pb^{2+} ion curve). The results of the experimental data treatment indicated the diffusion presence as being a preliminary stage in the sorption process of ions on PPT, though its role was small, especially in the case of Sr^{2+} ions. Thus, following the Lagergren model, the sorption of Pb^{2+} and Sr^{2+} ions was limited by the chemical ion exchange reaction stage, whereas the diffusion stage preceding ion-exchange sorption had a small additional impact on the reaction between sorbate and functional group of the sorbent.

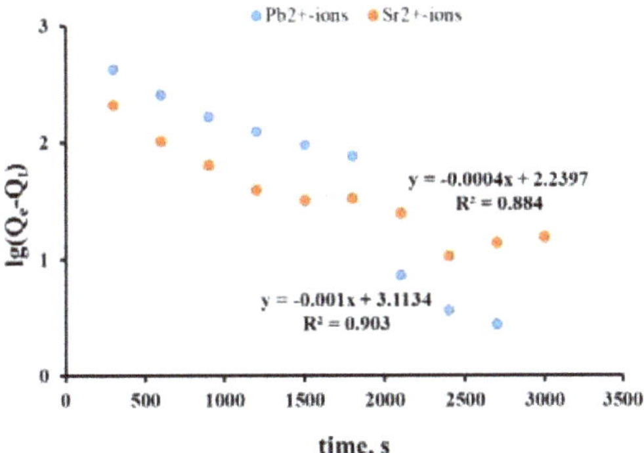

Figure 6. Sorption kinetics of Pb^{2+} and Sr^{2+} ions in the coordinates of the Lagergren pseudo-first-order.

The ongoing sorption processes more likely corresponded to the pseudo-second-order of Ho and McKay's model (Figure 7), regardless of the ion nature. The value of R^2 for the $\log(Q_e - Q_t)$ dependencies was 0.903 and 0.884 for Pb^{2+} and Sr^{2+} ions, respectively (Figure 6), compared to 0.971 and 0.997 for the t/Q_t dependence (Figure 7).

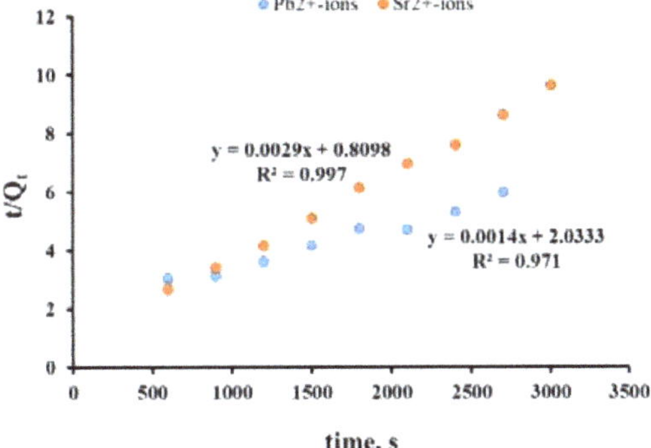

Figure 7. Sorption kinetics of Pb^{2+} and Sr^{2+} ions in the coordinates of the Ho and McKay pseudo-second-order.

Linear character of the dependences in the pseudo-second-order model's coordinates proved the limitation of ion sorption by the ion exchange stage; herein, this dependence was more defined for Sr^{2+}, and the reaction between the sorbent and the sorbate was a second-order reaction (they interacted with each other in one equivalent ratio). The linear curves allowed for the definition of the capacity and rate constant of the sorption (Table 2).

Table 2. Sorption kinetic parameters of Pb^{2+} and Sr^{2+} ions and PPT sorption capacity.

Ions	Rate Constants		PPT Sorption Capacity for Pseudo-Second-Order Model
	k, min^{-1}	k, $g \cdot mg^{-1} \cdot min^{-1}$	
	Pseudo–First Order Model	Pseudo–Second Order Model	
Pb^{2+}	0.0023	$9.6 \cdot 10^{-7}$	714.3
Sr^{2+}	0.0009	$10 \cdot 10^{-6}$	344.8

Analysis of the sorption kinetic parameters, namely, the pseudo-first- and pseudo-second-order sorption constants, as well as the sorption capacity, allowed us to find out the multistage interaction mechanism of PPT with metal ions. This was most likely explained by the large PPT specific surface area (90–100 m^2/g), as well as sufficiently large interlayer distance, which both caused the interaction of the sorbate on the outer surface with thiol groups (-TiOH) and on the inner surface by ionic substitution of potassium cations.

The pseudo-first-order model described indeed chemical sorption, as most probable chemical reactions including ion exchange and covalent bonding are confirmed by good compliance of experimental data. Herein, the assumption made when the chemical composition analysis of the PPT surface before and after sorption was proven by the ion exchange as a limiting stage of the metal cation absorption by PPT.

3.5. PPT Sorption Capacity Compared to Published Sorbents

On the basis of the calculated sorption capacity Q (mg of metal/g of sorbent) (Table 2) and the experimentally determined equilibrium time τ_{eq} (45 min, Figure 3), PPT sorption capacity exceeded the capacity of the most known titanium sorbents (Table 3).

Table 3. Reported adsorption capacities towards heavy metal ions by typical adsorbents.

Adsorbent	Ion	Adsorption Capacity, mg/g	Equilibrium Time, min	Adsorption Conditions	Reference
PPT	Pb^{2+}	714.3	45	pH = 4.5–6.2	This article
	Sr^{2+}	344.8		PH = 8.5–10.8	
Chitosan/TiO_2 hybrid adsorbent	Pb^{2+}	36.8		60 °C pH = 3–4	[38]
$Na_2Ti_3O_7$	Pb^{2+}	563.6	60–120	-	[39]
Titanates with various morphology	Pb^{2+}	105–304	90	-	[40]
Magnetic titanium nanotubes	Pb^{2+}	442.5	60	pH = 5	[41]
Titanate nanofibers	Pb^{2+}	244–280	-	pH = 6–7	[42]
	Sr^{2+}	50–55	-		
Titanate nanotubes	Pb^{2+}	2.6	-	pH = 5	[43]
		299.5	-		[44]
Titanate nanotubes obtained by hydrothermal method	Pb^{2+}	up to 2000	30	pH = 4	[45]
		520.8	180	pH = 5–6	[46]
	Sr^{2+}	91.7	10	-	[47]
		98.7	30	-	[48]

Thus, the sorption capacity of the synthesized PPT can exceed the capacity of most known literature titanium-containing sorbents with developed morphology of 1.3–270 times for Pb^{2+} and 3.5 times for Sr^{2+}.

4. Conclusions

In this study, X-ray amorphous potassium polytitanate sorbent with a layered structure and well-developed outer and inner surfaces were synthesized by a molten salt method, and its sorption characteristics were investigated. The sorption equilibrium during PPT aging in Pb^{2+} (pH< 5.5) and Sr^{2+} (pH< 8.5) solutions with a concentration of 50 g/L occurred in 45 min. The sorption capacity of PPT was calculated on the basis of a kinetic parameter study designed in accordance with pseudo-second order model and was equal to 714.3 mg/g and 344.8 mg/g for Pb^{2+} and Sr^{2+} ions, respectively.

The multi-stage interaction mechanism between ions and sorbent, including outer diffusion to sorbent surface, inner diffusion on sorbent surface, inner diffusion in the PPT interlayer space, and ion exchange chemical reaction where ion exchange occurs as a liming stage, was developed using Boyd's diffusion model and pseudo-first- and pseudo-second-order chemical kinetic models. Herein, Pb^{2+} sorption rate was significantly affected by the outer (within first minutes of interaction) and inner diffusion. Energy-dispersive analysis and quantum chemical study proved the lead and strontium presence in the PPT composition after sorption.

In summary, the sorption capacity of potassium polytitanate sorbent obtained in this study was significantly higher than the capacity of the most ceramic materials synthesized previously [38–47]. Despite better sorption of metal cation on carbon-based sorbents [15–17], ceramic materials are more preferable because of their re-use for various practical applications. For this reason, the synthesized material is a promising product for industry-scaled wastewater purification in practice.

Author Contributions: A.E.—Investigation; A.S.—Formal analysis; M.V.—Writing—original draft; T.B.—Visualization; S.A.—Validation; A.G. (Alexander Gorokhovsky)—Conceptualization; A.G. (Anna Godymchuk)—Writing—review & editing; I.B.—Project administration; P.O.O.—Writing—review & editing. All authors have read and agreed to the published version of the manuscript.

Funding: This research received no external funding.

Conflicts of Interest: The authors declare no conflict of interest.

References

1. Duruibe, J.O.; Ogwuegbu, M.O.C.; Egwurugwu, J.N. Heavy metal pollution and human biotoxic effects. *Int. J. Phys. Sci.* **2007**, *2*, 112–118.
2. Alvarez Ayuso, E.; Garcia Sánchez, A.; Querol, X. Purification of metal electroplating wastewaters using zeolites. *Water Res.* **2003**, *37*, 4855–4862. [CrossRef] [PubMed]
3. Tchounwou, P.B.; Yedjou, C.G.; Patlolla, A.K.; Sutton, D.J. Heavy metal toxicity and the environment. In *Molecular, Clinical and Environmental Toxicology*; Springer: Basel, Switzerland, 2012; pp. 133–164.
4. Chui, Z.C.; He, W.Q. *Industrial Wastewater Treatment*; Metallurgical Industry Press: Beijing, China, 1999.
5. Hu, Q.H.; Weng, J.Q.; Wang, J.S. Sources of anthropogenic radionuclides in the environment: A review. *J. Environ. Radioact.* **2010**, *101*, 426–437. [CrossRef] [PubMed]
6. Guhathakurta, H.; Kaviraj, A. Heavy metal concentration in water, sediment, shrimp (Penaeus monodon) and mullet (Liza parsia) in some brackish water ponds of Sunderban, India. *Mar. Pollut. Bull.* **2000**, *40*, 914–920. [CrossRef]
7. Ali, I.; Gupta, V.K. Advances in water treatment by adsorption technology. *Nat. Protoc.* **2006**, *1*, 2661. [CrossRef] [PubMed]
8. Jiuhui, Q.U. Research progress of novel adsorption processes in water purification: A review. *J. Environ. Sci.* **2008**, *20*, 1–13.
9. Shahmirzadi, M.A.A.; Hosseini, S.S.; Luo, J.; Ortiz, I. Significance, evolution and recent advances in adsorption technology, materials and processes for desalination, water softening and salt removal. *J. Environ. Manag.* **2018**, *215*, 324–344. [CrossRef]

10. Bhattacharyya, K.G.; Gupta, S.S. Adsorption of a few heavy metals on natural and modified kaolinite and montmorillonite: A review. *Adv. Colloid Interface Sci.* **2008**, *140*, 114–131. [CrossRef]
11. Płaza, A.; Kołodyńska, D.; Hałas, P.; Gęca, M.; Franus, M.; Hubicki, Z. The zeolite modified by chitosan as an adsorbent for environmental applications. *Adsorpt. Sci. Technol.* **2017**, *35*, 834–844. [CrossRef]
12. Karthikeyan, G.; Andal, N.M.; Anbalagan, K. Adsorption studies of iron (III) on chitin. *J. Chem. Sci.* **2005**, *117*, 663–672. [CrossRef]
13. Karthikeyan, G.; Anbalagan, K.; Andal, N.M. Adsorption dynamics and equilibrium studies of Zn (II) onto chitosan. *J. Chem. Sci.* **2004**, *116*, 119–127. [CrossRef]
14. Wu, F.-C.; Tseng, R.-L.; Juang, R.-S. A review and experimental verification of using chitosan and its derivatives as adsorbents for selected heavy metals. *J. Environ. Manag.* **2010**, *91*, 798–806. [CrossRef] [PubMed]
15. Rao, G.P.; Lu, C.; Su, F. Sorption of divalent metal ions from aqueous solution by carbon nanotubes: A review. *Sep. Purif. Technol.* **2007**, *58*, 224–231. [CrossRef]
16. Xue, C.; Qi, P.; Liu, Y. Adsorption of aquatic Cd^{2+} using a combination of bacteria and modified carbon fiber. *Adsorpt. Sci. Technol.* **2018**, *36*, 857–871. [CrossRef]
17. Guo, T.; Bulin, C.; Li, B.; Zhao, Z.; Yu, H.; Sun, H.; Ge, X.; Xing, R.; Zhang, B.; Zhang, B. Efficient removal of aqueous Pb(II) using partially reduced graphene oxide-Fe_3O_4. *Adsorpt. Sci. Technol.* **2018**, *36*, 1031–1048. [CrossRef]
18. O'Connell, D.W.; Birkinshaw, C.; O'Dwyer, T.F. Heavy metal adsorbents prepared from the modification of cellulose: A review. *Bioresour. Technol.* **2008**, *99*, 6709–6724. [CrossRef]
19. Ahalya, N.; Ramachandra, T.V.; Kanamadi, R.D. Biosorption of heavy metals. *Res. J. Chem. Environ.* **2008**, *7*, 71–79.
20. Hua, M.; Zhang, S.; Pan, B.; Zhang, W.; Lv, L.; Zhang, Q. Heavy metal removal from water/wastewater by nanosized metal oxides: A review. *J. Hazard. Mater.* **2012**, *211*, 317–331. [CrossRef]
21. Mishra, S.P.; Singh, V.K.; Tiwari, D. Radiotracer technique in adsorption study: Part XVII. Removal Behaviour of Alkali Metal (K- and Li-) Titanates for Cd(II). *Appl. Radiat. Isot.* **1998**, *49*, 1467–1475. [CrossRef]
22. Nunes, L.M.; Cardoso, V.A.; Airoldi, C. Layered titanates in alkaline, acidic and intercalated with 1,8-octyldiamine forms as ion-exchangers with divalent cobalt, nickel and copper cations. *Mater. Res. Bull.* **2006**, *41*, 1089–1096. [CrossRef]
23. Cardoso, V.D.A.; de Souza, A.G.; Sartoratto, P.P.; Nunes, L.M. The ionic exchange process of cobalt, nickel and copper (II) in alkaline and acid-layered titanates. *Colloids Surf. A* **2004**, *248*, 145–149. [CrossRef]
24. Bitonto, L.; Volpe, A.; Pagano, M.; Bagnuolo, G.; Mascolo, G.; La Parola, V.; Di Leo, P.; Pastore, C. Amorphous boron-doped sodium titanates hydrates: Efficient and reusable adsorbents for the removal of Pb^{2+} from water. *J. Hazard. Mater.* **2017**, *324*, 168–177. [CrossRef]
25. Sanchez-Monjaras, T.; Gorokhovsky, A.; Escalante-Garcia, J.I. Molten salt synthesis and characterization of potassium polytitanate ceramic precursors with varied TiO_2/K_2O molar ratios. *J. Am. Ceram. Soc.* **2008**, *91*, 3058–3065. [CrossRef]
26. Burmistrov, I.N.; Kuznetsov, D.V.; Yudin, A.G.; Muratov, D.S.; Milyaeva, S.I.; Kostitsyn, M.A.; Gorshenkov, M.V. Analysis of the Effect of Preparation Conditions for Potassium Polytitanates on Their Morphological Properties. *Refract. Ind. Ceram.* **2012**, *52*, 393–397. [CrossRef]
27. Tretyachenko, E.V.; Gorokhovsky, A.V.; Yurkov, G.Y.; Fedorov, F.S.; Vikulova, M.A.; Kovaleva, D.S.; Orozaliev, E.E. Adsorption and photo-catalytic properties of layered lepidocrocite-like quasi-amorphous compounds based on modified potassium polytitanates. *Particuology* **2014**, *17*, 22–28. [CrossRef]
28. Gorokhovsky, A.V.; Tret'yachenko, E.V.; Vikulova, M.A.; Kovaleva, D.S.; Yurkov, G.Y. Effect of Chemical Composition on the Photocatalytic Activity of Potassium Polytitanates Intercalated with Nickel Ions. *Russ. J. Appl. Chem.* **2013**, *86*, 343–350. [CrossRef]
29. Perdew, J.P.; Burke, K.; Ernzerhof, M. Generalized gradient approximation made simple. *Phys. Rev. Lett.* **1996**, *77*, 3865. [CrossRef]
30. Dyall, K.G. An exact separation of the spin-free and spin-dependent terms of the Dirac–Coulomb–Breit Hamiltonian. *J. Chem. Phys.* **1994**, *100*, 2118–2127. [CrossRef]
31. Visscher, L.; Dyall, K.G. Dirac-Fock atomic electronic structure calculations using different nuclear charge distributions. *At. Data Nucl. Data Tables* **1997**, *67*, 207–224. [CrossRef]

32. Lagergren, S. *Zur Theorie der Sogenannten Absorption Gelöster Stoffe*; P. A. Norstedt & Söner: Stockholm, Sweden, 1898.
33. Ho, Y.S.; Ng, J.C.Y.; McKay, G. Kinetics of pollutant sorption by biosorbents: Review. *Sep. Purif. Rev.* **2000**, *29*, 189–232. [CrossRef]
34. Biryukova, M.I.; Burmistrov, I.N.; Yurkov, G.Y.; Mazov, I.N.; Ashmarin, A.A.; Gorokhovskii, A.V.; Gryaznov, V.I.; Buznik, V.M. Development of a Fibrous Potassium Polytitanate. *Theor. Found. Chem. Eng.* **2015**, *49*, 485–489. [CrossRef]
35. Rabinovich, V.A.; Havin, Z.Y. *Handbook of Chemistry*; Khimiya: Leningrad, Russia, 1991; p. 432.
36. Zanello, P. *Inorganic Electrochemistry: Theory, Practice and Application*; Royal Society of Chemistry: London, UK, 2007.
37. Ivanov, V.M.; Figurovskaya, V.N.; Burmaa, D. Photometric determination of cobalt and erbium in their binary alloys. *Vestn. MGU Ser. 2 Khimiya* **1999**, *40*, 98–102.
38. Tao, Y.; Ye, L.; Pan, J.; Wang, Y.; Tang, B. Removal of Pb(II) from aqueous solution on chitosan/TiO_2 hybrid film. *J. Hazard. Mater.* **2009**, *161*, 718–722. [CrossRef] [PubMed]
39. Li, N.; Zhang, L.; Chen, Y.; Fang, M.; Zhang, J.; Wang, H. Highly efficient, irreversible and selective ion exchange property of layered titanate nanostructures. *Adv. Funct. Mater.* **2012**, *22*, 835–841. [CrossRef]
40. Huang, J.; Cao, Y.; Liu, Z.; Deng, Z.; Tang, F.; Wang, W. Efficient removal of heavy metal ions from water system by titanate nanoflowers. *Chem. Eng. J.* **2012**, *180*, 75–80. [CrossRef]
41. Zhang, L.; Wang, X.; Chen, H.; Jiang, F. Adsorption of Pb(II) Using Magnetic Titanate Nanotubes Prepared via Two-Step Hydrothermal Method. *CLEAN–Soil Air Water* **2014**, *42*, 947–955. [CrossRef]
42. Yang, D.; Zheng, Z.; Liu, H.; Zhu, H.; Ke, X.; Xu, Y.; Wu, D.; Sun, Y. Layered titanate nanofibers as efficient adsorbents for removal of toxic radioactive and heavy metal ions from water. *J. Phys. Chem. C* **2008**, *112*, 16275–16280. [CrossRef]
43. Liu, W.; Wang, T.; Borthwick, A.G.; Wang, Y.; Yin, X.; Li, X.; Ni, J. Adsorption of Pb^{2+}, Cd^{2+}, Cu^{2+} and Cr^{3+} onto titanate nanotubes: Competition and effect of inorganic ions. *Sci. Total Environ.* **2013**, *456*, 171–180. [CrossRef]
44. Wang, T.; Liu, W.; Xiong, L.; Xu, N.; Ni, J. Influence of pH, ionic strength and humic acid on competitive adsorption of Pb (II), Cd (II) and Cr (III) onto titanate nanotubes. *Chem. Eng. J.* **2013**, *215*, 366–374. [CrossRef]
45. Chen, Y.C.; Lo, S.L.; Kuo, J. Pb (II) adsorption capacity and behavior of titanate nanotubes made by microwave hydrothermal method. *Colloids Surf. A* **2010**, *361*, 126–131. [CrossRef]
46. Xiong, L.; Chen, C.; Chen, Q.; Ni, J. Adsorption of Pb (II) and Cd (II) from aqueous solutions using titanate nanotubes prepared via hydrothermal method. *J. Hazard. Mater.* **2011**, *189*, 741–748. [CrossRef] [PubMed]
47. Ryu, J.; Kim, S.; Hong, H.-J.; Hong, J.; Kim, M.; Ryu, T.; Park, I.-S.; Chung, K.-S.; Jang, J.; Kim, B.-G. Strontium ion (Sr^{2+}) separation from seawater by hydrothermally structured titanate nanotubes: Removal vs. Recovery. *Chem. Eng. J.* **2016**, *304*, 503–510. [CrossRef]
48. Wen, T.; Zhao, Z.; Shen, C.; Li, J.; Tan, X.; Zeb, A.; Wang, X.; Xu, A.-W. Multifunctional flexible free-standing titanate nanobelt membranes as efficient sorbents for the removal of radioactive $^{90}Sr^{2+}$ and $^{137}Cs^{+}$ ions and oils. *Sci. Rep.* **2016**, *6*, 20920. [CrossRef] [PubMed]